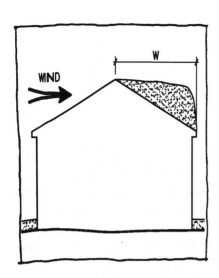

D0621850

Snow Loads

About the Authors

Michael O'Rourke, Ph.D, P.E., received his B.S. in Civil Engineering from Illinois Institute of Technology and his Masters and Ph.D. from Northwestern. During most of his 29 years on the faculty in Civil Engineering at Rensselaer Polytechnic Institute, he has been involved in snow load research sponsored by the U.S. Army Cold Regions Research and Engineering Lab, National Bureau of Standards, National Science Foundation, and the Metal Building Manufacturers Association, among others. This research work has resulted in publication of roughly two dozen referred journal papers and conference proceedings. Dr. O'Rourke has been a member of the ASCE 7 Snow and Rain Loads Committee since 1978 and has been Chair since 1997.

Peter D. Wrenn, P.E., received his BS and Masters in Civil Engineering from Rensselaer Polytechnic Institute. He is an Associate Structural Engineer with the Quality Assurance Unit for the Dormitory Authority–State of New York (DASNY). Prior to joining DASNY, he specialized in the structural rehabilitation and restoration of historic buildings, working on a number of national landmark structures throughout the United States. He was first involved with snow load research as an undergraduate assistant. His graduate work involved the similitude of backward stepped, flat roof snow loads utilizing a water flume. He has participated in snow-related structural failure investigations, case-study design snow load analysis at the U.S. Army Cold Region Research and Engineering Lab, and snow load research for the Metal Building Manufacturers Association.

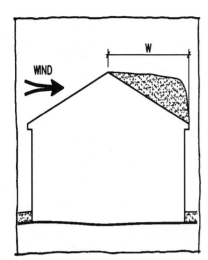

Snow Loads
A Guide to the Use and Understanding of the Snow Load Provisions of ASCE 7-02

Michael O'Rourke
Peter D. Wrenn

Cataloging-in-publication data on file with the Library of Congress.

Authors' Disclaimer

Although the authors have done their best to ensure that any advice, recommendation, interpretation, or information given herein is accurate, no liability or responsibility of any kind (including liability for negligence) is accepted by the authors.

Published by the American Society of Civil Engineers
1801 Alexander Bell Drive
Reston, VA 20191
www.asce.org

Contents

List of Figures

List of Tables

Preface

This guide provides practicing structural engineers with a detailed description of the snow load provisions of SEI/ASCE Standard 7-02, *Minimum Design Loads for Buildings and Other Structures*, published by the American Society of Civil Engineers (ASCE). The intent of this guide is to present the research and philosophy that underpins the provisions and to illustrate the application of the provisions through numerous examples. Readers and users of this guide will know how to use the provisions and also the reasoning behind the provisions. In this fashion, users may be able to address non-routine snow loading issues that are not explicitly covered in the Standard. Every effort has been made to make the illustrative example problems in this guide correct and accurate. The authors welcome comments regarding inaccuracies, errors, or different interpretations. The views expressed and the interpretation of the snow load provisions made in this guide are those of the authors and not of the ASCE 7 Standards Committee or the ASCE organization.

Acknowledgments

The authors would like to acknowledge the past and present members of the Snow and Rain Loads Committee of ASCE 7. Without their comments, questions, and discussions, the development of Section 7 in SEI/ASCE Standard 7, and subsequently this Guide, would not have been possible.

As with any document of this type, many individuals have contributed their hard work and effort, other than the authors listed. The authors acknowledge the work and effort extended by the administrative staff of the Department of Civil and Environmental Engineering at Rensselaer Polytechnic Institute, who assisted in the word processing and preparation of the narrative. The authors also would like to acknowledge the sketch work prepared by Christopher Keado, AIA, who graciously contributed the hand-drawn illustrations associated with each chapter.

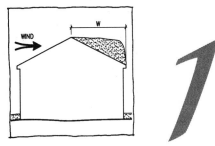

Introduction

Roof snow loading must be considered for all states in the United States, with the exception of Florida, which has a mapped ground snow load of zero. Snow is the *controlling* roof load, over wind or roof live load, in roughly half of the states. Specifically, when the ground snow load is 20 psf or greater, snow loading typically controls for at least some roof structural components.

Snow loading is a frequent and costly cause of structural performance problems, including collapse. For example, the March 1993 East Coast storm, which cost an estimated $1.75 billion, is one of the ten worst natural catastrophes in the United States in terms of insurance claims paid. As shown in **Table I-1**, this 20-state storm, also called the "Blizzard of the Century," was more costly than the Oakland Hills fire, Hurricanes Fran and Iniki, and the Loma Prieta earthquake. Although the winter of 1992–1993 was a record-setting year in terms of losses due to snow, a large insurer reports that the following winter (1993–1994) resulted in roughly 120 roof snow and ice load losses at a total cost of about $100 million. Hence, snow loading and snow load provisions are something with which structural engineers involved in building design need to be familiar.

Lightweight roof framing systems are sensitive to snow overload. The ASCE 7-02 Commentary notes the increased importance of snow overload as the live-to-dead load ratio increases. Consider the case of a 25-psf design snow load and a 15-psf snow overload. If the dead load is 50 psf (live-to-dead load ratio of 25/50 = 0.5), the 15-psf snow overload corresponds to a 20% overload in terms of the total load (90/75 = 1.20). If, on the other hand, the dead load is 5 psf (live-to-dead load ratio of 25/5 = 5.0), the 15-psf snow overload would correspond to a 50% overload in terms of total load (45/30 = 1.50).

Table I-1   Ten Costliest U.S. Natural Catastrophes

Month/Year	Catastrophe	Estimated Insured Loss ($M)
August 1992	Hurricane Andrew	15,500
January 1994	Northridge earthquake	12,500
September 1989	Hurricane Hugo	4,195
October 1995	Hurricane Opal	2,100
March 1993	"20-State winter storm"	1,750
October 1991	Oakland Hills fire	1,700
September 1996	Hurricane Fran	1,600
September 1992	Hurricane Iniki	1,600
May 1995	Southwest flooding	1,135
October 1989	Loma Prieta earthquake	960

Source: National Research Council 1999.

Such differences become apparent when loss information is reviewed. As an example, a series of mixed precipitation events (snow, ice, and rain) resulted in structural damage to more than 1,600 man-made facilities in the Pacific Northwest during the 1996–1997 holiday season. Detailed information for a subset 88 structures is available in a 1998 report by the Structural Engineers Association of Washington (SEAW). For that subset, the roof systems most frequently damaged were flat, wood panelized roofs (36 out of 88); wood trusses including short span, long span, and bowstring (18 out of 88); metal building systems (7 out of 88), and wood girder-joist systems (7 out of 88). The absence from the SEAW subset list of comparatively heavy reinforced or prestressed concrete roof systems is not surprising. It is unusual to see a snow-related collapse of flat plate, flat slab, one-way joist, or other types of concrete systems.

Table I-2 presents the snow load type that was the primary cause of partial or complete collapse of more than 40 buildings. **Table I-2** suggests that roof step and parapet wall drifts accounted for about 30% of the failures, with gable roof drift loading (unbalanced loading due to across-the-ridge drifting) accounting for another 30%. A little less than 10% of the failures were due to nominally uniform loading on freezer buildings and cold rooms wherein the indoor temperature was intentionally kept at or below freezing. The final category, "other," includes the case of a partial collapse due to ice dam formation at the eave of a building, a few cases with missing flange braces in metal buildings, and a number of cases where the overload was due to a combination of effects. A typical example of combined effects is a gable roof with an east-west ridgeline, which abuts a tall structure located immediately to the west. Wind out of the southeast would result in

Table I-2 Types of Snow Loading Resulting in Damage

Snow Load Type	Number of Cases*	%
Roof step snow drift	7	17
Parapet wall snow drift	5	12
Across-the-ridge gable drift	12	29
Nominally uniform load on "freezer" buildings	3	7
Other	14	35
Total	41	100

*Cases are from O'Rourke's forensic engineering practice over the past decade.

drifting loads at the lower roof's northwest corner. This lower level roof drift is partially due to windward roof step drifting (wind component out of the east) in combination with gable roof drifting (wind component out of the south). In **Table I-2,** such a case of combined loading is classified as "other."

The purpose of this guide is to provide practicing structural engineers with a detailed description of the snow loading provisions in Section 7.0 of the SEI/ASCE Standard 7-02, *Minimum Design Loads for Buildings and Other Structures.* The guide presents the research and philosophy that underpins the provisions and illustrates the application of the provisions through numerous examples so that the user not only knows how to use the provisions, but also knows the reasoning behind them. In this fashion, users may be able to address nonroutine snow loading issues that are not explicitly covered in the ASCE 7-02 provisions.

Practicing structural engineers involved in the design, analysis, and/or review of building structures are the intended audience for this guide. Although not necessary, it is expected that readers have used the ASCE 7 snow load provisions in the past, even if only in a "cookbook" fashion. This guide could be used in graduate or undergraduate civil engineering courses, but it was not written with that audience exclusively in mind.

Chapters 2 through 10 of this guide are named and numbered to correspond with the ASCE 7-02 snow provisions. As an example, Chapter 5 is devoted to partial loading, which is covered in Section 7.5 of ASCE 7-02. This guide does not contain a separate chapter for Section 7.1 because the symbols and notations provided in that section are defined in this guide when first encountered. The other exception is Chapter 11, which corresponds to material in Sections 7.11 and 7.12 of ASCE 7-02. Chapter 12 answers Frequently Asked Questions (FAQs).

The examples in Chapters 2 through 10 illustrate the correct application of the ASCE 7-02 snow provisions. Chapter 11 provides background information on the provisions in Sections 7.11 and 7.12 of ASCE 7-02, and Chapter 12 offers views, thoughts, and ideas on snow loading issues that are *not* addressed in the Standard. As such, the methods and approaches presented in Chapter 12 are offered as guidance and do not constitute an official interpretation of ASCE 7-02.

For ease of reference, the Appendix contains snow load provisions of ASCE 7-02. Frequent users of the snow provisions may choose to read the guide chapters while occasionally referring to the actual code in the Appendix. Readers who are not familiar with the snow provisions in ASCE 7-02 should read the corresponding code section in the appendix before beginning a new guide chapter.

A note about figures, tables, and equations presented in this guide: All the figures, tables, and equations of Section 7.0 of ASCE 7-02 are identified with Arabic numbers (e.g., Figure 7-2, Table 7-4, Eq. (7-3)). To avoid confusion, the figures, tables, and equations that are unique to this guide are identified with Roman numbers in bold print (e.g., **Figure III-2, Table VI-3, Eq. (X-1)**).

2
Ground Snow Load

The roof snow load provisions in ASCE 7-02 are based on or related to the ground snow load, p_g. This approach, which mirrors Canadian practice, is used because of the relative abundance of ground snow measurement information in comparison to roof snow load measurements. As described in more detail in the ASCE 7-02 Commentary, the ground snow map (Figure 7-1) is based on concurrent recordings of ground load and depth at 204 National Weather Service (NWS-water equivalent) stations in combination with about 9,200 ground depth sites from the Soil Conservation Service (SCS), the National Weather Service (NWS-depth), and other recording agencies.

Since ground snow depth, of and by itself, is of little interest to structural engineers, a clever method was used to establish an equivalent snow density or unit weight for the SCS and NWS-depth data. Specifically, the NWS-water equivalent data (which included both the load and the depth) was used to generate a relationship between the 50-yr ground snow depth and the 50-yr ground snow load, which in turn was applied to the depth-only data (i.e., SCS). The relationship (Tobiasson and Greatorex 1996) between 50-yr ground snow load (p_g, in psf) and 50-yr ground snow depth (h_g, in inches) is

$$p_g = 0.279 h_g^{1.36}$$

(Eq. II-1)

This relationship is plotted in **Figure II-1**. In essence, **Eq. (II-1)** establishes an equivalent density or unit weight. For example, the equivalent density is about 8 pcf for a snow depth of 1 ft (i.e., 50-yr depth of 12 in. corresponds to a 50-yr load of 8.2 psf) and about 12 pcf at 3 ft (i.e., 50-yr depth of 36 in. corresponds to a 50-yr load of 36.5 psf). This nonlinear increase is due par-

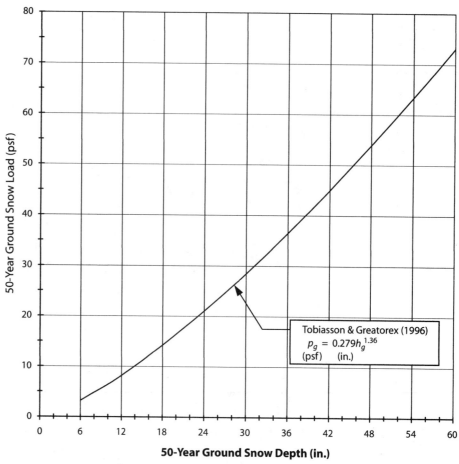

Figure II-1 Relationship between 50-yr Ground Snow Load and 50-yr Ground Snow Depth Used in Determining ASCE 7-02 Snow Load Map

tially to the self-weight of the snow that compacts the snow toward the bottom of the snowpack. A comparison of **Eq. (II-1)** with other density relations is presented in Chapter 7.

The ground snow load map for the United States (Figure 7-1) presents the 50-yr mean recurrence interval (MRI) ground snow. That is, the ground snow load has a 2% annual probability of being exceeded. For most of the Central Midwest, bounded by Indiana on the east and Nebraska on the west, the ground snow loads are simply a function of latitude. As one might expect, Louisiana has relatively small ground snow loads (0 psf or 5 psf) whereas Wisconsin has relatively large values (25 psf to 70 psf).

2.1 Influence of Latitude, Elevation, and Coastlines

In the eastern United States, the ground snow load, p_g, generally increases with latitude, but two additional variables also influence p_g: the distance from the coastline, which affects ground snow loads from Virginia to New England, and site elevation. An example of the coastal effect is shown in

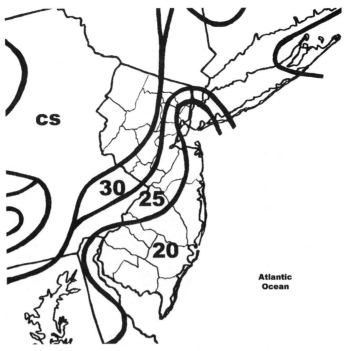

Figure II-2   Portion of ASCE 7-02 Ground Snow Load Map Showing Increase in p_g with Distance from Shoreline in New Jersey

Figure II-2 for New Jersey. At a given latitude, ground snow loads typically increase as one moves inland or away from the coastline. Elevation is also a factor in the East because of the string of mountains along the Appalachian Trail. In some locations, such as eastern Tennessee or Rochester, New York, the mapped ground snow load value in Figure 7-1 (10 psf and 40 psf, respectively) applies to sites with elevations less than the given upper elevation limit (1,800 ft and 1,000 ft, respectively). That is, designers are provided with ground snow load information at lower elevations where most of the buildings are located. At elevations greater than the upper limits, a site-specific case study (CS) is required.

Latitude and elevation also influence ground snow load values in the West. For instance, the 50-yr ground snow load at a given elevation in New Mexico is typically less than that for the same elevation in Montana. However, because of the more rugged and variable terrain, the overall pattern in the West is more complex. Unlike most regions in the Midwest and some in the East, where the 50-yr ground snow load is strictly a function of latitude, all the ground snow loads in the West are a function of site elevation. For some locations, such as southeast Arizona, ground snow loads are specified for a range of elevations: zero for elevations of 3,500 ft or less, 5 psf for elevations between 3,500 ft and 4,600 ft, etc. Other locales, such as the majority of western Colorado, require site-specific case studies.

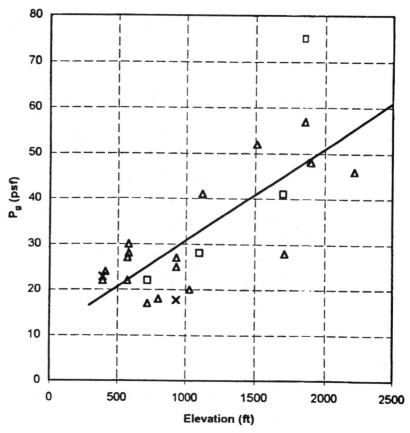

Figure II-3 Case Study Plot of 50-yr Ground Snow Load versus Elevation for Sites Near Freedland, Pa.

2.2 Site-Specific Case Studies

All locations represented with a "CS" on Figure 7-1 require a site-specific case study in order to establish the design ground snow load. As noted on the map in relation to CS areas, "the extreme local variations in ground snow loads in these areas preclude mapping at this scale." Also, at all sites that have a higher elevation than that designated on the map in parentheses, the ground snow load must be established by a case study. For example, a case study is required for all areas in eastern Tennessee that have an elevation higher than 1,800 ft. As described in more detail by Tobiasson and Greatorex (1996), a case study involves regressing 50-yr ground snow load values for a number of sites, in close proximity to the site of interest, versus elevation. The least squares straight line then establishes the local "reverse lapse" rate which in turn can be used to establish the 50-yr ground snow load for the site of interest. The lapse rate is the decrease in temperature for a unit increase in elevation. As used herein, a "reverse lapse" rate is the increase in ground snow load for a unit increase in elevation.

A case study prepared by the U.S. Army Cold Regions Research and Engineers Lab (CRREL) for Freedland, Pa., is shown in **Figure II-3.** Note that there were 23 sites within a 25-mi radius of Freedland with known values of p_g. When plotted versus elevation, the least squares line has an approximate slope of 2 psf per 100 ft of elevation difference. From the plot, the 50-yr ground snow load for Freedland with an elevation of 1,880 ft was 48 psf. CRREL currently provides site-specific case studies that are similar to the Freedland case study shown in **Figure II-3** on request and free of charge.

The ASCE 7-02 Commentary also contains a town-by-town listing of ground snow loads for New Hampshire. Note in this regard that a "town" in New Hampshire is a relatively large political subdivision of a county and does not simply denote a small city or large village. These values, developed by Tobiasson et al. (2002) using the case study approach outlined above, are for a specific elevation within each town. For different site elevations within a town, it was determined that a reverse lapse rate of 2.1 psf per 100 ft is appropriate for the entire state of New Hampshire.

Note that the reverse lapse rate is not uniform across the United States. That is, although the rate of 2.1 psf per 100 ft for New Hampshire is essentially identical to the 2.0 psf per 100 ft from **Figure II-3** for Freedland, Pa., it is substantially larger than that for many locations in the West. For example, in north central Arizona, the ground snow load is 5 psf for sites between 3,000 and 4,500 ft, 10 psf for 4,500 to 5,400 ft, and 15 psf for 5,400 to 6,300 ft. This corresponds to a reverse lapse rate of about 0.5 psf per 100 ft of elevation difference, or about 25% of the New Hampshire rate.

The ASCE 7-02 Commentary also refers to documents with valuable snow load information for Arizona, Colorado, Idaho, Montana, Oregon, Washington, and parts of California. These references, typically prepared by a state structural engineers association or a state university, present 50-yr ground snow loads; some also present snow provisions. Note that for some locations, the 50-yr ground load value using a state reference is different than that from the ASCE 7-02 map. For example, the most recent state map for Washington (Structural Engineers Association of Washington (SEAW) 1995) contains isolines that, when multiplied by the site elevation, give the 50-yr ground snow load. For a site in Bellingham, Wash., with an elevation of 100 ft, the SEAW procedure gives a 50-yr ground load of 15 psf, while the ASCE 7-02 map gives 20 psf. Unfortunately, the ASCE 7-02 text and Commentary are not clear about which 50-yr value is to be used: the ASCE 7-02 value, the state value, the larger, the average, or other permutation. It is anticipated that future versions of ASCE 7 will recommend using the state or local ground snow load map in such cases, as long as the state/local document meets certain criteria (e.g., 50-yr values are based on an extreme value statistical analysis). Nevertheless, the ASCE 7-02 text is clear that a 50-yr MRI ground snow load is to be used in its provisions. Furthermore, despite the exact source of the ground snow load value, it also is clear that a design according to the ASCE 7-02 provisions requires the use of the ASCE exposure and thermal factors, drifting relations, etc., as opposed to alternate provisions that may be part of the aforementioned state documents.

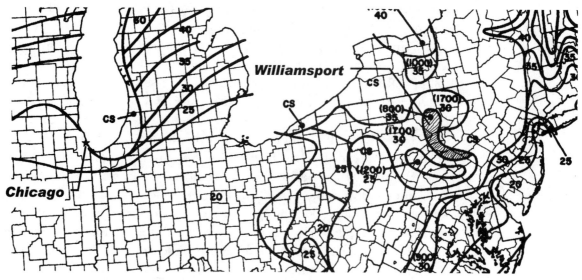

Figure II-4   Portion of ASCE 7-02 Ground Snow Load Map Showing Chicago, Ill., and Williamsport, Pa. (Zone for Williamsport is Cross-Hatched)

2.3   Example 2.1: Ground Snow Loads

Determine the 50-yr ground snow load for (a) Chicago, Ill., (b) Williamsport, Pa., and (c) a site in Enfield, N.H., with an elevation of 1,150 ft.

Solution

a) Chicago, on the southwestern shore of Lake Michigan, is in a 25-psf ground snow load zone, as shown in **Figure II-4.**

b) Williamsport, located in north central Pennsylvania, has a ground elevation of 528 ft as per the TopoZone.com web site. Furthermore, Williamsport is located in the cross-hatched region of **Figure II-4** where ground snow load of 35 psf is given for an elevation ≤800 ft. Since the case-specific site elevation of 528 ft is less than the 800 ft upper elevation, a case study is not required, and the 50-yr ground snow load is 35 psf.

c) As shown in Table C7-4 of the ASCE 7-02 Commentary, the 50-yr MRI ground snow load for Enfield, N.H., is 85 psf at the reference elevation of 1,300 ft. The site elevation is less than the reference elevation. Therefore, the reverse lapse rate of 2.1 psf per 100 ft and the elevation difference of 150 ft are used to determine the ground snow load:

$$p_g = 85 \text{ psf} - \frac{2.1 \text{ psf}}{100 \text{ ft}}(1,300 \text{ ft} - 1,150 \text{ ft})$$
$$= 82 \text{ psf}$$

3

Flat Roof Snow Loads

Section 7.3 in ASCE 7-02 provides the calculations for the flat roof snow load, p_f, and prescribes a minimum roof snow load. The flat roof snow load formula incorporates effects of wind exposure of the site, thermal conditions of the roof system, and the importance of the facility. The flat roof snow load is combined with the slope factor, C_s, from Section 7.4 to form a design snow load for which all roofs must be checked. The minimum roof load is intended to cover situations where the "typical" wind and thermal effects are not applicable.

3.1   Measured Conversion Factors

Case histories have shown that the snow load on a roof is typically less than that on the ground provided drifting is absent. To establish a rational basis for roof snow loads, the Cold Regions Research and Engineers Lab (CRREL) sponsored a program in the late 1970s in which university researchers made simultaneous measurements of the ground and roof snow loads. The researchers measured a group of structures with varied wind and thermal environments over the course of a few winters, and O'Rourke, Koch, and Redfield (1983) subsequently analyzed the data.

Figure III-1 shows the measured loads for one structure over the course of the 1976–1977 winter. For this structure, the ground and roof loads were nominally the same on the three sampling dates in December; however, the ground load was four or five times greater than the roof load for both the mid-February and early-March samples. Although there are a number of values for the ratio of simultaneous roof to ground loads (ranging from about 1.0 prior to January to about 0.23 in February and March),

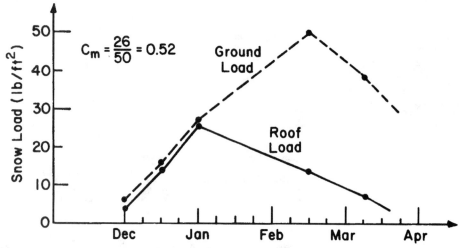

Figure III-1 Sample Variation of Ground and Roof Snow Loads for One Structure during 1976–1977 Winter

the ratio of the maximum roof load (26 psf on or about New Year's Day) to the maximum ground load (50 psf in mid-February) is of most interest to structural engineers. The ratio of maximum roof to maximum ground load, C_m, is 0.52 (i.e., 26/50), although they did not occur on the same day. Henceforth, this ratio is referred to as the ground-to-roof conversion factor, or simply the conversion factor. Note that the maximum annual roof load is obtained through multiplication of the maximum annual ground load and this conversion factor.

The region's weather patterns affect the conversion factor. In mild locations where the winter is characterized by a few snowfalls separated by warmer weather, there will be little or no accumulation of ground snow from one snowfall to another. For such locations, both the maximum ground load and the maximum roof load tend to occur immediately after the largest snowfall. Hence, C_m would be relatively close to 1.0. On the other hand, consider a colder location where the winter is characterized by a greater number of snowfalls that are closely spaced. Both the maximum ground and roof snow loads are caused by the snow's accumulation and melting. Wind, thermal, and other effects modify the maximum roof load further. For such locations, the conversion factors typically range from 0.3 to 1.0. Hence, areas with infrequent snowfalls and small accumulations tend to have higher ground-to-roof conversion factors than colder areas with substantial ground snow accumulation.

As stated earlier, the roof's exposure to wind and its thermal characteristics influence the conversion factors. In terms of wind exposure, the CRREL study roofs were characterized as being sheltered, semisheltered, or windswept. As one moves from sheltered structures to windswept structures, the conversion factor goes down. Examples of all three exposure classifications are shown in **Figures III-2** through **III-4**. In relation to thermal characteristics, structures were classified as heated or unheated.

Figure III-2   Example of Sheltered Roof in CRREL Study

Figure III-3   Example of Semisheltered Roof in CRREL Study

Figure III-4   Example of Windswept Roof in CRREL Study

Figure III-5 shows the influence of wind exposure rating on the conversion factor. If the data is normalized by the middle rating (i.e., semisheltered), the conversion factors range from 1.3 (i.e., 0.79/0.59) to 0.9 (i.e., 0.53/0.59). **Figure III-6** presents similar information for the roof's thermal rating. Normalizing by the heated category, the conversion factors range from 1.27 to 1.0.

Table III-1 presents the conversion factor data for the study subdivided by both thermal and exposure ratings. For example, the 12 unheated structures classified as having a semisheltered wind exposure had an average conversion factor of 0.66.

3.2   Flat Roof Snow Load

In ASCE 7-02, the flat roof snow load for design purposes is

$$p_f = 0.7 C_e C_t I p_g \tag{Eq. 7-1}$$

where

C_e = Dimensionless ASCE 7-02 exposure factor
C_t = Dimensionless ASCE 7-02 thermal factor
I = Dimensionless importance factor
p_g = 50-yr ground snow load discussed in Chapter 2 of this guide

The "0.7" factor is intended to represent a somewhat conservative average conversion factor for a "typical" roof with $C_e = C_t = I = 1.0$. As we will soon see, a heated structure of ordinary importance in a partially exposed, suburban site would have $C_e = C_t = I = 1.0$. Assuming that "partially exposed"

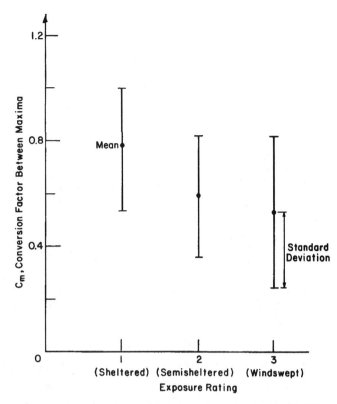

Figure III-5 Conversion Factor Mean and Standard Deviation for Three Exposure Ratings

Table III-1 Average Conversion Factors from CRREL Study

Exposure Rating	Heated	Unheated
Windswept	0.52 (18)	0.55 (14)
Semisheltered	0.48 (8)	0.66 (12)
Sheltered	0.66 (8)	0.84 (17)

Note: Value in parentheses represents the number of roofs in each subcategory.

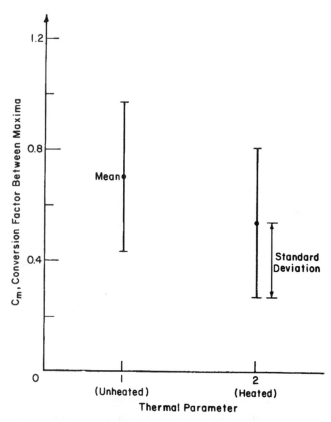

Figure III-6 Conversion Factor Mean and Standard Deviation for Two Thermal Ratings

is defined essentially the same as "semisheltered," **Table III-1** suggests that the true "average" C_m would be 0.48. However, the somewhat conservative 0.7 factor is used because of the scatter observed in the study described above.

3.3   Exposure Factor (C_e)

The ASCE 7-02 exposure factors are shown in Table 7-2. They are a function of the surface roughness category (ranging from city center to shoreline) and the location of the structure within the regional terrain (ranging from fully exposed to sheltered). The surface roughness categories are defined in Section 6.0 (Wind Loads). They are intended to capture the overall windiness of the area surrounding the site and define the variation of wind velocity with height within the atmospheric boundary layer. The variation of the wind speed with height is due to the friction at the earth's surface caused by obstructions such as buildings and trees (the higher the obstruction, the larger the friction). The wind slows as it gets closer to the ground and is often modeled by the so-called Power Law:

$$\frac{V(z)}{G} = \left(\frac{z}{\delta}\right)^{\alpha}$$

(Eq. III-1)

Table III-2 Power Law Coefficients from Davenport (1965)

Terrain	α Power Law Coefficient	δ Boundary Layer Thickness (ft)
Open Terrain	0.16	900
Suburban Terrain	0.28	1,300
Centers of Large Cities	0.40	1,700

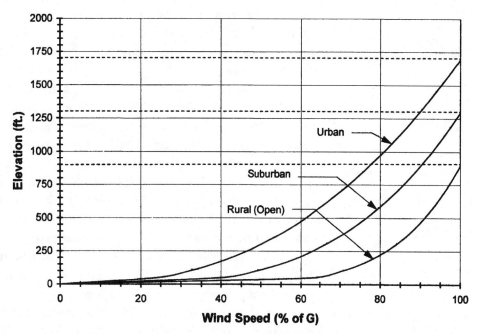

Figure III-7 Variation of Wind Speed with Elevation for Three Terrain Categories and Constant Geostropic Speed

where

$V(z)$ = Wind velocity at elevation z
G = Geostropic wind speed at the top of the atmospheric boundary layer
δ = Thickness of the atmospheric boundary layer
α = Power Law coefficient

Values for δ and α as suggested by Davenport (1965) are shown in **Table III-2.** The variation of wind speed with height for Davenport's three terrain classes is shown for a constant geostropic speed in **Figure III-7.** As one might expect, the thickest boundary layer is associated with the terrain that has the most surface friction. Also for the same geostropic speed, the wind velocity at a given roof elevation would be largest for open terrain and smallest for city centers. This is why the seashore is a great place to fly a kite.

The decreased values for C_e in Table 7-2 as one moves from Terrain Category A (city center) to Terrain Category D (shoreline) reflect the tendency for more snow to be blown off roofs with more wind. The same trend applies in terms of the local exposure classification. Within a specific terrain category, the wind experienced at the roof level decreases as the exposure changes from fully exposed to sheltered, and the tabulated values for C_e increase correspondingly.

The exposure factors for design purposes range from 0.7 to 1.3 as shown in Table 7-2. This range is broader than the 0.9 to 1.3 range for the "normalized" conversion factors from **Figure III-5.** Note, however, that the $C_e = 0.7$ value in Table 7-2 is for fully exposed structures at treeless or "above the tree line in windswept mountainous" terrain. Such terrain was not represented in the CRREL study. Hence, one could argue that Table 7-2 is reasonably consistent with the conversion factor measurements determined in the CRREL study.

Footnotes at the bottom of Table 7-2 provide definitions of the various roof exposures used in ASCE 7-02. At one extreme is "sheltered," which corresponds to roofs tight against conifers that qualify as obstructions. The other extreme is "fully exposed," which corresponds to a roof with no obstructions, including large rooftop equipment and tall parapet walls (parapets that extend above the height of the balanced snow). The middle class, "partially exposed," is arguably the most common roof exposure. It corresponds to all roofs that are not "sheltered" or "fully exposed." It should be noted that two roofs on the same structure could have different roof exposures. A classic example is a two-level roof in which the upper level roof is "fully exposed" while the lower level is "partially exposed" due to the obstruction provided by the upper level roof.

The footnotes for Table 7-2 also provide a definition for obstructions that provide shelter. Specifically, if the top of an object such as an adjacent building or a group of trees is "h_o" above the roof elevation, then it must be located within a distance "$10\,h_o$" of the roof to be considered an obstruction that provides shelter. In relation to these sorts of sheltering effects, Sachs (1972) presents useful data. At the transition from wooded terrain to more open terrain, the following approximate reductions in wind velocity were noted: 60% reduction in velocity at 5 tree heights downwind, 45% reduction at 10 tree heights, and ranging from 10% to 30% at 20 tree heights. A rule of thumb, often used by sailors and windsurfers, is that you need to be six tree heights downwind in order to get "good wind." Based on the above, the "$10\,h_o$" requirement in ASCE 7-02 is reasonable.

Although Table 7-2 provides an obstruction distance of $10\,h_o$, it does not establish where the measurement is taken. For example, if a potential obstruction is located north of a roof, it is unclear if $10\,h_o$ is measured from the building's north wall, south wall, or other reference point. When faced with such a situation in practice, the authors suggest measuring from the roof's geometric center.

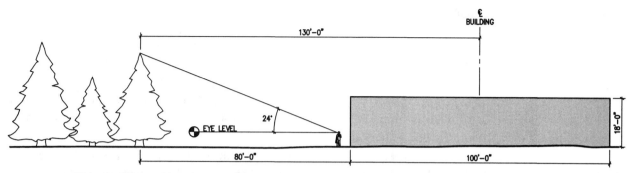

Figure III-8 North Elevation of Proposed Building Located in a Clearing for Ex. 3.1

3.4   Example 3.1: Roof Exposure

A new building is to be sited west of a large clearing in a forest. The mean roof height is 18 ft. As shown in **Figure III-8,** the distance from the edge of the trees to the building's east wall is 80 ft and the east-west building dimension is 100 ft. Using an Abney level at the eastern edge of the building, the treetops are located at 24 degrees up from the horizontal. Do these trees serve as an obstruction that provides shelter with respect to the center of the roof?

Solution

From trigonometry, the tops of the trees are 80 tan 24° = 35.6 ft above eye level, which is assumed as 6.0 ft. Hence, the tops of the trees are nominally 42.0 ft above ground level. Since the roof elevation is 18.0 ft, the elevation difference from the roof to the top of trees equals 24.0 ft, or h_o = 24.0 ft.

As per the problem statement, the point of reference is the center of the roof, which is 130 ft from the edge of the clearing. Since 130 ft is less than 10 h_o (240 ft), the trees serve as obstructions that provide shelter with respect to the center of the roof. If all sides of the building are similarly sheltered, then the building's roof would be categorized as "sheltered." If not, the roof would be categorized as "partially exposed."

In Ex. 3.1, an alternate approach is possible if the distance between the building and the edge of the clearing (i.e., the 80-ft dimension in **Figure III-8**) is difficult to determine (for instance, due to the presence of a swamp). In that case, one could sight the top of the trees from both the east and west ends of the building (angles from the horizontal of 24.0 and 11.2 degrees, respectively) and apply trigonometry to determine the distance between the building and the edge of the clearing.

3.5   Thermal Factor (C_t)

The ASCE 7-02 thermal factors are shown in Table 7-3. They range from 0.85 for specified greenhouses to 1.2 for unheated structures and structures intentionally kept below freezing. A C_t factor of 1.1 is assigned to structures kept just above freezing and to certain well-insulated "cold roofs." A cold roof is one in which air passageways allow sufficient inflow at the eaves and outflow through ridge vents.

The intent of using thermal factors is to quantify the differences in ground-to-roof conversion factors due to heat loss through the roof layer. By driving through a suburban area four or five days after a significant snowfall, one can see this effect. The roof snow on the heated homes will be noticeably less than the roof snow present on the unheated garages.

The C_t values in Table 7-3 suggest that the roof snow load atop an unheated structure would typically be about 20% greater than for a heated structure, whereas the observed conversion factor data in **Figure III-6** suggests a slightly larger value of 27% (i.e., unheated mean/heated mean $0.70/0.55 = 1.27$).

It should be noted that for unheated structures of ordinary importance ($C_t = 1.2$ and $I = 1.0$), the flat roof snow load, as given in Eq. (7-1), could be larger than the ground snow load. That is, for C_e of 1.2 and 1.3, the flat roof design loads equal $1.01\ p_g$ and $1.09\ p_g$, respectively. At first glance, it seems odd that the roof snow load would be larger than the ground snow load absent drifting or sliding. Nevertheless, a number of case history observations demonstrated this exact phenomenon.

One such set of case histories came from the 1996–1997 holiday storms in the Pacific Northwest. A 1998 report by the Structural Engineers Association of Washington (SEAW) presented measured ground and roof snow load data for the Yakima area. The annual maximum ground snow was reported to be 32 psf on December 29, 1996, whereas the annual maximum flat roof snow load atop a number of freezers and cold rooms was reportedly equal to 36 psf. Thus, the ground-to-roof conversion factor for the 1996–1997 winter was 1.13 for these facilities. In general, all the air below the insulated roof layer at these facilities was kept at or below freezing. These structures did not simply house a freezer unit; they were, in essence, large freezers. It should be noted that snow drifting was not a significant factor at the facilities. Therefore, although the same total amount of precipitation fell on the cold room roofs as on the ground, the roofs accumulated more snow because the snow on the roof melted at a slower rate between storms than the ground snow. For heated structures absent drifting and sliding, the roof load never exceeds the ground snow load.

Sack et al. (1984) presents another case history with a conversion factor greater than 1.0. Periodic ground and roof snow measurements were made over a 2-yr period at a U.S. Forest Service facility in McCall, Id. The structure was located at a relatively sheltered site, and a roof covered a loading dock area on one side of the building. Since the loading dock was not enclosed, the bottom of the "overhang" roof was open to ambient air temperatures, which, on average, were below freezing. The observed conversion factors for the overhang roof were 1.27 and 1.28 for the 1982–1983 and 1983–1984 winters, respectively. As explained by Sack, the comparatively large conversion factors for the overhang are due, at least in part, to effects of the ground heat flux. In early winter, particularly before frost has set in, the earth's surface is comparatively warm and heat flux from the earth melts some of the ground snow. The roof snow located on the overhang, however, is not subject to the ground heat flux since it has ambient air directly below.

3.6   Importance Factor (*I*)

The importance factor, *I*, is used to determine all the environmental live loads (i.e., flood, wind, snow, and earthquake). Its purpose is to increase environmental live loads for structures that are particularly important, and to allow a reduction for structures that are not. The various categories of structures are identified in Table 1-1 and are based on the nature of the building's occupancy or intended use. Structures are deemed important when the potential for loss of human life is particularly high (e.g., elementary schools with capacity greater than 150 and explosives manufacturing facilities are both in Category III), or if they are essential in time of disaster (e.g., emergency shelters, fire stations, and hospitals are in Category IV). Unimportant structures are those where the potential for loss of human life is particularly low (e.g., minor storage facilities and agricultural facilities are in Category I). Ordinary structures (Category II) are by definition those not in Categories I, III, or IV. Note that a structure's "importance," as defined in ASCE 7-02, is unrelated to its initial cost, replacement cost, economic value, or the value of its contents. A structure's importance relates, directly or indirectly, to the hazard potential to human life in the event of failure.

The mapped values in Figure 7-1, the map of 50-yr ground snow loads, are intended for ordinary structures (Category II; $I = 1.0$). For structures in Category I, the *I* factor is 0.8, which corresponds to a 25-yr ground snow load. As noted in the Commentary, the average value of the ratio of 25-yr to 50-yr ground snow loads for more than 200 sites across the United States is 0.8. For structures in Category IV, the *I* factor is 1.2, which corresponds to a 100-yr ground snow load. Hence, for snow loading, the return period used in design is a function of the relative importance of the structure, based on its intended use and the nature of occupancy.

3.7   Minimum Roof Loads

In summary, the flat roof snow load (as well as the sloped roof load, which is discussed in Chapter 4 of this guide) incorporates the generalized ground-to-roof conversion factor of 0.7, as well as the C_e and C_t factors, which increase or decrease the roof snow load depending on the specific wind exposure and thermal environment. Except for some special circumstances discussed below, this results in a roof snow load that is less than the ground snow load.

In low ground snow load areas, the region's design ground snow load could be the result of a single large storm. If the winds were calm during and after this single large storm, the roof load would be comparable to the ground snow load because the wind and thermal effects would not have had sufficient time to significantly alter the roof snow load. For instance, in areas with design ground snow loads (reference Figure 7-1) roughly in the 5-psf to 20-psf range, it is possible that a single storm could result in *both* the ground and roof having equivalent loads approaching the 50-yr design snow load, p_g. In areas with larger design ground snow loads (p_g in Figure 7-1 of 25 psf or more), it is still possible for a single large storm to result in ground and roof loads being equivalent; however, it is unlikely that these loads would approach the 50-yr design load. Expecting loads greater than 20 psf from a single large storm is unreasonable.

The minimum roof snow load (defined in Section 7.3) is needed for this reason. The minimum roof snow load is a function of the ground snow load, p_g, and the importance factor. Specifically, when p_g is comparatively low (20 psf or less), a single large storm may result in a roof load approaching p_g, and the minimum roof load is defined as the importance factor multiplied by p_g. On the other hand, when p_g is substantial (more than 20 psf), a single large storm is unlikely to result in a roof load of p_g, and the minimum load is defined as 20 psf multiplied by the importance factor.

The roof geometry influences whether the minimum roof load applies. Consider, for example, a gable roof. If the roof slope is shallow or low, and as a result unbalanced or across-the-ridge drift loads (see Section 7.6) do not apply, then the minimum roof load may well control the design. On the other hand, if the gable roof slope is moderate to steep (i.e., across-the-ridge drift loading is applicable), then the Section 7.3.4 minimum load is unlikely to control the design. In this case, an analysis involving the minimum roof load could arguably be performed by inspection.

The minimum load provisions outlined in ASCE 7-02 reflect this logic. Minimum roof loads *are* required for monoslope roofs with slopes less than 15 degrees; for hip and gable roofs with slopes (in degrees) less than or equal to $(70/W) + 0.5$ (where W is the horizontal distance from eave to ridge in feet); and for curved roofs where the secant angle from eave to crown is less than 10 degrees. As shown in Chapter 6, unbalanced loads are *not* required for hip and gable roofs with slopes less than $(70/W) + 0.5$ and for curved roofs with a secant slope less than 10 degrees. Hence, if unbalanced loading is not required, then minimum roof loads from Section 7.3.4 must be considered during the design.

As noted previously, all building roofs are required to be designed for a uniform snow load. The controlling uniform load will be either the sloped roof snow load or the minimum roof load discussed above.

For certain low-sloped roofs located in low ground snow load areas, a 5-psf rain-on-snow surcharge must be added to the flat roof snow load. This rain-on-snow surcharge applies in regions with $p_g \leq 20$ psf and to roofs with slopes less than ½ on 12 (approximately 2.4 degrees). Chapter 10 provides examples and further discussion about rain-on-snow surcharges.

4
Sloped Roof Snow Load

Section 7.4 of ASCE 7-02 provides information needed to convert the flat roof snow loads, p_f, into sloped roof snow loads, p_s, using the roof slope factor, C_s. As mentioned previously, the sloped roof snow load is the basis for determining the snow load for all structures. For certain low-sloped roofs in low ground load areas, a rain-on-snow surcharge is added to p_s (refer to Chapter 10 of this guide); and the sloped roof snow load without surcharge serves as the balanced load for partial loading as well as various drift and sliding load situations discussed in subsequent chapters.

As shown in Eq. (7-2), the sloped roof snow load, p_s, is the product of the roof slope factor, C_s, and the flat roof snow load, p_f:

$$p_s = C_s \, p_f \hspace{4cm} \text{(Eq. 7-2)}$$

As one might expect, the slope factor is a decreasing function of the roof slope. As skiers know, it is difficult to "keep" snow on steep trails. For roofs, the reduction is caused by snow sliding off after initially being deposited or simply not remaining on the roof surface in the first place.

This sliding after initial deposition can initiate at the snow–roof surface interface. Sliding at the interface is facilitated by a combination of slippery roof surfaces and lubrication resulting from snow melting on poorly insulated, "warm" roofs. Hence, C_s is lower for both slippery and warm roof surfaces.

Snow sliding also can occur *above* the interface, within the roof snow layer itself. Snow will slide off a cold roof with a nonslippery surface if the slope is steep enough. This is the same mechanism as a typical avalanche, which initiates at a weak layer *within* a snowpack. Of course, obstructions on the roof—such as large or multiple vent pipes, snow guards, or eave ice dams—can inhibit the snow from sliding. In addition, a sufficient space below the eave, where the sliding snow may be deposited, must remain unobstructed in order for snow to slide off the roof. For instance, the ASCE 7-02 Commentary cites that A-frame roofs should not be classified as unobstructed because their eaves extend close to the ground.

Graphs of C_s versus roof slope for three thermal factor values are provided in Figure 7-2. As indicated above, C_s is a decreasing function of roof slope and an increasing function of C_t. Two curves are shown for each value of C_t: the dashed lines are for unobstructed slippery surfaces, and the solid lines are for all other surfaces (i.e., nonslippery surfaces and obstructed, slippery surfaces). Again, unobstructed slippery surfaces (dashed lines) have smaller values of C_s, and all other surfaces (solid lines) have larger values of C_s for a given thermal factor. To avoid confusion, Section 7.4 provides examples of roof surfaces that are considered slippery (e.g., metal, slate, and plastic membrane) as well as nonslippery (e.g., asphalt and wood shingles and shakes).

When $C_t \leq 1.0$ (i.e., a warm roof), the roof must meet an additional requirement in order for the surface to be considered unobstructed and slippery. Specifically, for unventilated roofs, the R-value must be greater than or equal to 30 °F hr ft^2/BTU; for ventilated roofs, the R-value must be greater than or equal to 20 °F hr ft^2/BTU. This R-value requirement is intended to ensure that the roof remains unobstructed from eave ice dams. Ice dams tend to form on poorly insulated, warm roofs and can hinder the sliding of roof snow.

The roof slope factors presented in Figure 7-2 are reasonably consistent with available case history information. This is illustrated in **Figure IV-1,** which plots data collected from a 1988 study by Sack with values prescribed by the ASCE 7-02 code. Arguably, Sack's study provides the best information on slope effects. Over a 4-yr period in the early 1980s, Sack and his graduate students measured roof and ground snow load on more than a dozen roofs of various slopes in the McCall, Id., area. All the roofs in the slope study were unheated (i.e., C_t = 1.2) and slippery. The general area was wooded (Terrain Category B), and in terms of exposure, "The wind speed at all of the observation sites during the winter is nominal." This statement regarding wind speed suggests that the exposure factor from Table 7-2 would be either C_e = 1.0 (if partially exposed) or C_e = 1.2 (if sheltered). **Figure IV-1** presents the Idaho conversion factor data plotted versus roof slope. As expected, the conversion factor decreases as the roof slope increases. **Figure IV-1** also shows the conversion factor calculated as per the ASCE 7-02 procedures with an importance factor of 1.0.

$$(C_m)_{code} = 0.7 C_e C_t C_s \qquad \textbf{(Eq. IV-1)}$$

where

C_e = Exposure factor = either C_e = 1.0 or C_e = 1.2

C_t = Thermal factor = 1.2

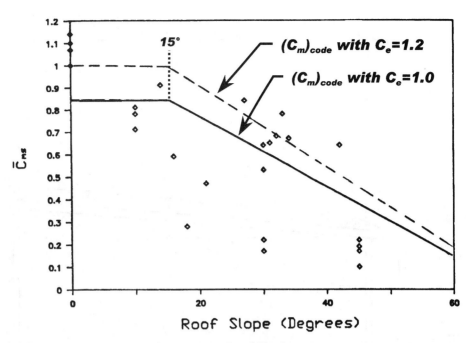

Figure IV-1 Idaho Conversion Factor Data versus Roof Slope

C_s = Value as per Figure 7-2(c) with unobstructed slippery surfaces

The code conversion factor $(C_m)_{code}$ with $C_e = 1.2$ envelopes most, but not all, of the data points. In general, the variation of observed conversion factors with slope is consistent with the C_s relations in ASCE 7-02.

4.1 Example 4.1: Uniform Snow Load, Monoslope Roof (1 on 12)

Determine the uniform design roof snow loading for an emergency vehicle storage garage located in suburban Cleveland, Oh. (see **Figure IV-2**). The interior temperature in the winter is kept at about 38 °F to avoid frozen water pipe damage. The garage is located on a relatively open site about 30 ft from a taller municipal building that serves as an obstruction (i.e., the municipal building roof is at least 3 ft taller than the garage's). The garage has a monoslope roof with slope of 1 on 12.

Solution

The monoslope roof is classified as partially exposed, due to the presence of the taller municipal building; because the locale is suburban (Terrain Category B), the C_e factor is 1.0. The thermal factor, C_t, is 1.1 since the structure is kept just above freezing. Although the structure is occupied by vehicles most of the time, it still falls in Category IV of Table 1-1, and the importance factor from Table 7-4 is 1.2. For the Cleveland area, $p_g = 20$ psf from Figure 7-1. Hence, the flat roof snow load is

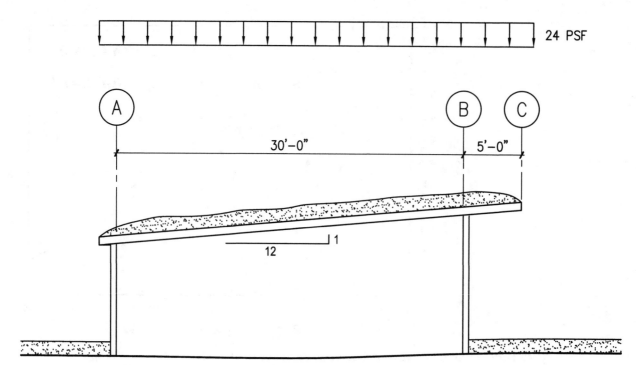

Figure IV-2   Uniform Design Load and Elevation of a Monoslope Roof for Ex. 4.1

$$p_f = 0.7 C_e C_t I p_g$$
$$= 0.7(1.0)(1.1)(1.2)(20)$$
$$= 18.5 \text{ psf}$$

From Figure 7-2, for cold roofs with $C_t = 1.1$, $C_s = 1.0$ for a 1-on-12 roof slope (4.76 degrees) with any surface classification. Hence, the sloped roof snow load is equivalent to flat roof load, 18.5 psf. Because the roof slope is 1 on 12, which is greater than ½ on 12, a rain-on-snow surcharge as per Section 7.10 need not be considered.

Since the monoslope roof has a slope less than 15 degrees (4.76° < 15°), the minimum roof load must be considered. For this case with $p_g = 20$ psf and $I = 1.2$, the minimum roof load becomes (I) (p_g) or 24 psf. Hence, the uniform design load is 24 psf as shown in **Figure IV-2.**

Refer to Ex. 5.2 for the partial snow loading of this roof structure.

4.2 Example 4.2: Uniform Snow Load, Monoslope Roof (4 on 12)

Same as Ex. 4.1, except the monoslope roof is 4 on 12 and has an unobstructed, slippery roof surface.

Solution

From Ex. 4.1, the flat roof snow load is 18.5 psf, with an importance factor I = 1.2 and a thermal factor C_t = 1.1. From the equations in the ASCE 7-02 Commentary for a 4-on-12 (18.4 degree slope), unobstructed, slippery roof with C_t = 1.1,

$$C_s = 1.0 - (18.4° - 10°)/60° = 0.86$$

Hence, the sloped roof snow load is

$$p_s = C_s p_f$$
$$= 0.86 (18.5 \text{ psf})$$
$$= 15.9 \text{ psf (round to 16 psf)}$$

For this comparatively steep slope, there is no need to consider a rain-on-snow surcharge. Also, since the monoslope is steeper than 15 degrees, there is no need to consider minimum loads. Hence, the uniform design load is 16 psf.

4.3 Example 4.3: Uniform Snow Load, Wide Gable Roof

Determine the design uniform snow load for the unheated structure of ordinary importance shown in **Figure IV-3.** The site is in a suburban area (Terrain Category B) with a few nearby trees less than 10 h_o from the structure that are tall enough to be considered obstructions. The ground snow load for the area is 30 psf.

Solution

For a partially exposed roof in Terrain Category B, C_e = 1.0. Because the structure is unheated, C_t = 1.2 (Table 7-3). The structure is of ordinary importance (Category II); therefore, I = 1.0 from Table 7-4. C_s = 1.0, irrespective of surface slipperiness for ½-on-12 (2.39 degrees) slope; therefore, the sloped roof snow load is

$$p_s = 0.7 C_e C_t C_s I p_g$$
$$= 0.7(1.0)(1.2)(1.0)(1.0)(30 \text{ psf})$$
$$= 25 \text{ psf}$$

The rain-on-snow surcharge does not need to be considered because the slope is equal to ½ on 12 and the ground snow is greater than 20 psf. Also, the minimum loading as per Section 7.3.4 does not need to be considered because the slope is greater than $(70/W)$ + 0.5 [(70/130) + 0.5 = 1.04°]. Hence, the design uniform loading is 25 psf as shown in **Figure IV-3.**

The unbalanced load for this structure is determined in Ex. 6.2.

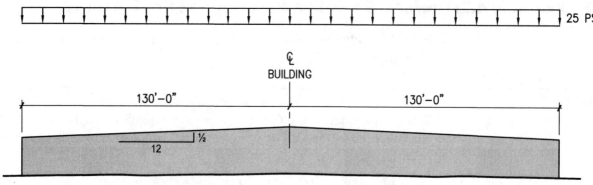

25 PSF

Ç
BUILDING

130'–0"   130'–0"

$\frac{1}{2}$
12

Figure IV-3   Uniform Design Load and Elevation of a Gable Roof for Ex. 4.3

4.4   Example 4.4: Uniform Snow Load, Narrow Gable Roof

Same as Ex. 4.3, except the eave-to-ridge distance is only 30.0 ft as shown in **Figure IV-4.**

Solution

From Ex. 4.3, p_g = 30 psf, I = 1.0, and p_s = 25 psf. There is no need to consider the rain-on-snow surcharge because p_g > 20 psf and the sloped roof uniform design load is 25 psf.

The gable roof slope of ½ on 12 (2.39 degrees) is less than $70/W + 0.5$ [$(70/30) + 0.5 = 2.83°$]; thus, minimum roof load requirements from Section 7.3.4 must be considered. Since p_g > 20 psf, the minimum roof load becomes $(I)(20) = 1.0(20) = 20$ psf. The minimum roof load is less than 25 psf; hence, the sloped roof uniform load controls, as shown in **Figure IV-4.**

In Ex. 4.4, the eave-to-ridge distances and the roof slopes are small enough that unbalanced loads (see Chapter 6) are not considered. If the roof framing were continuous, partial loading (see Chapter 5) would need to be considered. However, these nonuniform loads would be separate load cases.

Balanced loads for circular arc and sawtooth-type roofs are determined in Ex. 6.3 and Ex. 6.4.

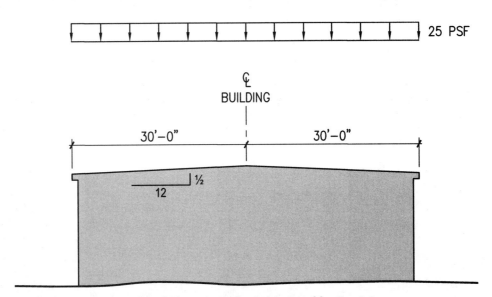

Figure IV-4   Uniform Design Load and Elevation of a Gable Roof for Ex. 4.4

5
Partial Loading

The partial loading provisions of Section 7.5 of ASCE 7-02 take into account patterned or checkerboard loading. The Commentary (reference Section C7.5) explains that partial loading is required for continuous structural members, such as roof purlins in metal building systems, where a *reduction* in snow loading on one span results in an *increase* in stress or deflection in an adjacent span. The Standard requires the designer to apply half of the full-balanced snow load to the partially loaded spans and the full-balanced snow load to the remainder of the spans. Note that partial loads are different from unbalanced or drift loads, which are discussed in Chapter 6. Unbalanced and drift loading arise when snow is removed from one portion of the roof and accumulates on another portion. With partial loading, snow is removed from one portion of the roof (for example, via wind scour or thermal effects), but it is not added to another portion.

The full-balanced snow load described in Section 7.5 corresponds to that given in Eq. (7-2) (i.e., $p_s = C_s P_f = 0.7 C_e C_t C_s I p_g$). The minimum roof load requirements defined in Section 7.3.4 are not applicable in the partial load provisions. The minimum roof loads are based on a single snowfall event with no allotted time for wind or thermal effects. This situation is different from that accounted for in the partial loading provisions, in which a reduction in snow load over time is envisioned for a portion of the roof due to wind, thermal, or other similar actions. Similarly, as noted in the Commentary in Section C7.10, the rain-on-snow provisions only need to be considered for a uniform or balanced load case, not for partial loading.

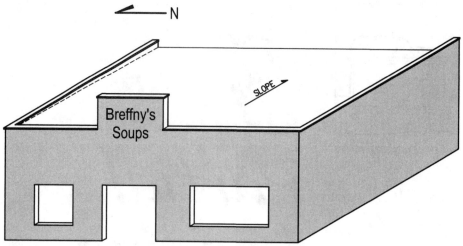

Figure V-1   West Elevation of a Building with a Pediment. Wind out of the West Results in Nonuniform Loading

5.1   Continuous-Beam Systems

Although partial snow loading for a roof can be compared to the checker-board loading pattern used for floor design, an actual checkerboard pattern is *not* prescribed for roof snow. Instead, continuous-beam systems are investigated for the effects of three loading cases as described in Section 7.5.1 and as illustrated in Figure 7-4 of the Standard. Case 1 may occur when two separate snow events, both corresponding to half the balanced design load, are separated by an intervening event—such as sleet, freezing rain, or crust formation—which prevents the lower half of the roof snow-pack from drifting. The second snowfall is then followed by a strong wind blowing from right to left. As it blows across the roof from right to left, it removes snow from all of the spans while simultaneously depositing snow on all of the downwind spans. Therefore, there is no net change for any span with an upwind source of driftable snow, yet there is a net snow removal for all of the downwind spans, which are without an upwind source. Before the driftable snow is blown off the last downwind span, the wind stops leaving a pattern corresponding to Case 1.

The pattern in Case 2 could be analogous to Case 1 except the wind event is shorter in duration and is traveling from left to right. In Case 2, the wind event is only long enough to clean the driftable snow (half the balanced load in the scenario under discussion) from the first upwind span.

The pattern in Case 3 could occur on the structure shown in **Figure V-1**. The structure has a monoslope roof (sloping toward the east at ½ on 12) without a parapet and with a pediment located near the center of the west wall. Since the pediment would be an obstruction, one would classify the whole roof as partially exposed and use the appropriate C_e factor from Table 7-2 to establish the balanced snow load for the entire roof. For a strong wind from the west, snow would be removed from the roof via wind scour for the north and south portions of the roof; however, little to no wind scour would occur directly behind the pediment, leaving a pattern similar to Case 3.

Note that the prescribed cases for the investigation of continuous-beam systems do not cover all permutations or patterns of partial roof loading (e.g., the Breffny's Soups sign in **Figure V-1** could correspond to three spans as opposed to the two spans presented in Case 3). The representative cases were chosen to cover situations that could reasonably be expected to occur without burdening the designer with numerous partial load cases, which, although conceivable, are unlikely to govern the design. For example, Cases 1 and 2 specifically target the end span which would have the largest midspan bending moment given equal span lengths and uniform load from span to span. Finally, partial load provisions are *not* required for structural members (e.g., a frame girder in metal building systems) that span perpendicular to the ridgeline in gable roofs (with slopes in degrees greater than or equal to $70/W + 0.5$). As we will see later, such gable roofs need to be designed for unbalanced loads due to across-the-ridge drifting. Although it is conceivable that there could be true partial loading on structural components spanning from eave to eave (as shown in **Figure V-2(a)**), the more common distribution is an unbalanced or an across-the-ridge drift load (as shown in **Figure V-2(b)**). Since a true partial loading as shown in **Figure V-2(a)** is uncommon and an unbalanced loading likely governs, the designer is not burdened with a partial load check for those members.

5.2   Other Structural Systems

Continuous-beam type components such as purlins and frames in metal building systems are statically indeterminate. For other types of structural systems, Section 7.5.2 requires that the partial load (half the balanced load) that produces the greatest effect on the member being analyzed must be chosen. A cantilevered roof girder system with "drop-in" simply supported spans, as shown in **Figure V-3,** is an example of a statically determinate system that must be analyzed for partial loading.

For the exterior girder (continuous over one support) and the interior girder (continuous over both supports), two partial load cases are investigated. In both cases, the cantilever and the adjoining drop-in span are considered as one region, while the portion of the cantilevered girder between the supports is considered to be another region.

In **Figure V-4,** Case A maximizes the moment, shear, and deflection for the region between the girder supports, whereas Case B, shown in **Figure V-5,** maximizes the same quantities in the cantilever and drop-in span. Note that a partial load (i.e., variations of load between the two links at either end) is not required for the drop-in span because the drop-in span is simply supported.

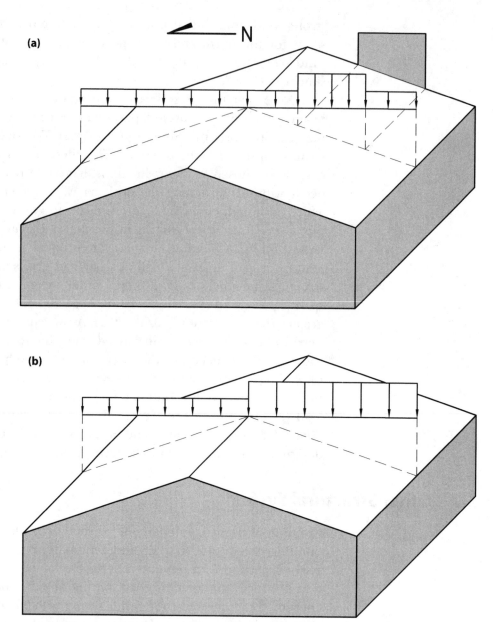

(a)

N

(b)

Figure V-2   West Elevation of a Gable Roof Structure. (a) Wind from the East Results in Partial Loading. (b) Wind from the North Results in Across-the-Ridge Drifting

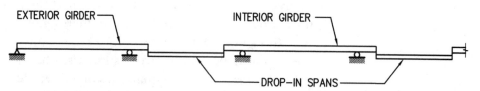

EXTERIOR GIRDER

INTERIOR GIRDER

DROP—IN SPANS

Figure V-3   Example of a Cantilevered Roof Girder System

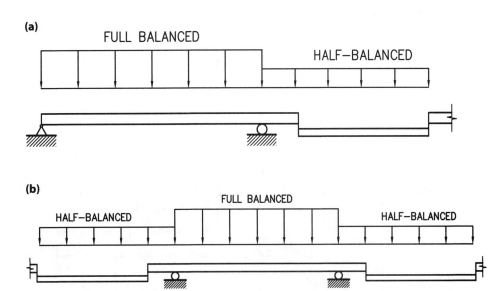

Figure V-4 Partial Load Case "A" for a Cantilevered Roof Girder System. (a) Exterior Girder. (b) Interior Girder.

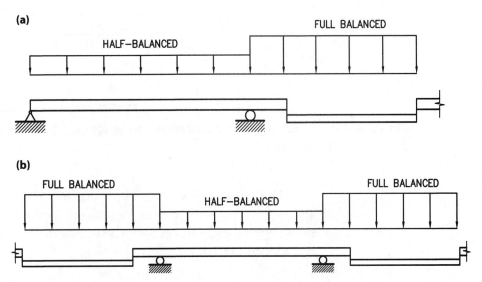

Figure V-5 Partial Load Case "B" for a Cantilevered Roof Girder System. (a) Exterior Girder. (b) Interior Girder.

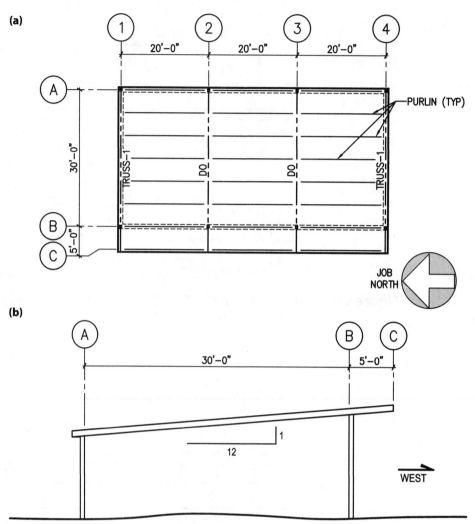

(a)

PURLIN (TYP)

TRUSS-1 DO DO TRUSS-1

20'-0" 20'-0" 20'-0"

30'-0"

5'-0"

JOB NORTH

(b)

30'-0" 5'-0"

1
12

WEST

Figure V-6 Emergency Vehicle Storage Garage for Ex. 5.1. (a) Roof Framing Plan. (b) North Elevation.

The ASCE 7-02 Commentary also mentions snow removal operations and melting as possible causes of partial loading. The issue of appropriate snow removal procedures for structures with continuous-beam components is discussed in Chapter 12.

5.3   Example 5.1: Uniform and Partial Snow Load, Monoslope Roof with Overhang

Determine the snow loading for the purlins and wood roof trusses in the emergency vehicle storage garage outlined in Ex. 4.1 (see **Figure V-6**). The purlins are 5 ft on center and span 20 ft. The purlins are fastened to and are supported by the wood trusses with metal hangers (simple supports). The wood roof trusses, in turn, span 30 ft 0 in. and cantilever 5 ft 0 in. beyond the front wall.

Solution
From Ex. 4.1, the sloped roof snow load p_s = 18.5 psf, while the minimum roof load is 24 psf.

Purlins: Since the purlins are simply supported, partial loading is not a consideration and the minimum roof load of 24 psf controls. Hence, for 5-ft purlin spacing, the controlling uniform roof snow load for the purlins is 5.0 ft × 24 psf = 120 lb/ft.

Roof Trusses: For uniform snow loads, the minimum roof load controls (24 psf > 18.5 psf). Therefore, the uniform load on the roof trusses is 20 ft × 24 psf = 480 lb/ft. Because the roof truss is continuous over the front wall at column line B, partial loading also must be considered.

For partial loading provisions, the balanced load is given by Eq. (7-2) (p_s = $C_s p_f$ = 0.7 $C_e C_t C_s I p_g$) or 18.5 psf in this case. For the 20-ft spacing, the two partial loads are

$$\text{balanced} = 20 \text{ ft} \times 18.5 \text{ psf} = 370 \text{ lb/ft}$$

$$\text{half-balanced} = ½ (370 \text{ lb/ft}) = 185 \text{ lb/ft}$$

The resulting three roof snow load cases, one uniform and two partial, to be evaluated are presented in **Figure V-7.**

Either the uniform load case or partial load Case 2 will control shear, moment, and deflection between A and B. The uniform case will control shear and moment in the cantilever. Either the uniform or partial load Case 1 will control the deflection at the end of the cantilever.

Since it is a cold roof, the eave ice dam provisions of Section 7.4.5 do not apply. Finally, since the separation distance is more than 20 ft, snow drifting from the taller municipal building on to the lower garage need not be considered. Ex. 7.3 presents a case in which such drifting is considered.

5.4   Example 5.2: Partial Snow Load, Continuous Purlins in Gable Roof (1 on 12)

Given the same conditions outlined in Ex. 4.1 for an emergency vehicle storage garage, determine the partial loading on the purlins for the gable roof building shown in **Figure V-8.**

Solution
The conditions are the same as those given in Ex. 4.1, so both the flat and sloped roof load is equal to 18.5 psf and the minimum roof load is equal to 24 psf. The problem statement does not ask for the controlling *uniform* load on the purlins since that likely will be governed by unbalanced (across-the-ridge) drifting, which is covered in Chapter 6. Nor does the problem statement ask for partial loading on the *frames* since, again due to unbalanced loading, partial loads are not required for structural elements that run per-

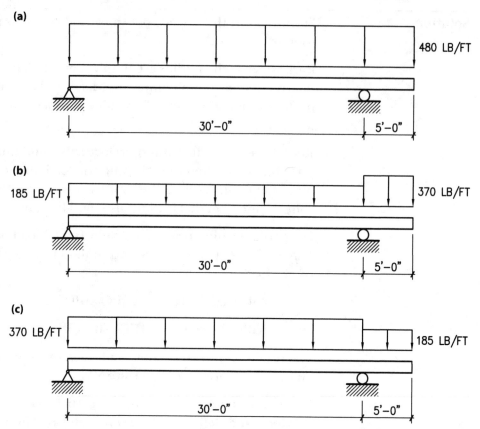

Figure V-7 Uniform and Partial Load Cases for Roof Trusses in Ex. 5.1. (a) Uniform Loading. (b) Partial Load Case 1. (c) Partial Load Case 2.

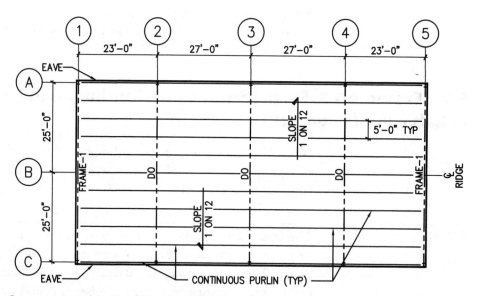

Figure V-8 Symmetric Gable Roof Framing Plan for Ex. 5.2

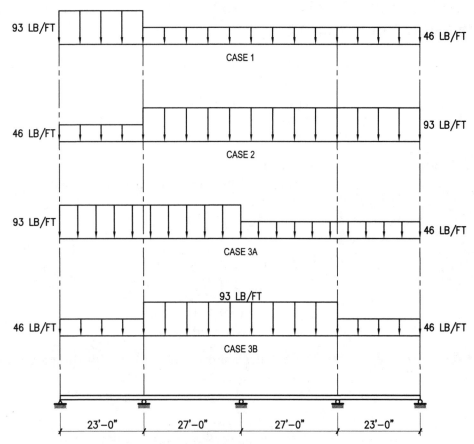

Figure V-9 Partial Load Cases for Roof Purlins in Ex. 5.2

pendicular to the ridge. That is, since the roof slope of 4.76 degrees (1 on 12) is larger than $70/W + 0.5 = [70/25 + 0.5 = 3.3°]$, partial loading on the continuous frames does not need to be investigated.

For partial loading, the balanced load is given by Eq. (7-2) (18.5 psf in this case), and the tributary width is 5.0 ft. Therefore,

full-balanced = 18.5 psf × 5.0 ft = 92.5; round to 93 lb/ft

half-balanced = ½(92.5 lb/ft) = 46.3 lb/ft; round to 46 lb/ft

The resulting partial load cases are shown in **Figure V-9.** Case 3 has three possible loading distributions, but only two are shown in **Figure V-9** (Cases 3A and 3B). Because of the roof plan symmetry, the third Case 3 distribution (46 lb/ft over the two leftmost spans, and 93 lb/ft on the others) is already covered in Case 3A.

5.5 Example 5.3: Partial Snow Load, Continuous Purlins in Gable Roof (3 on 12)

Determine the partial loading on the purlins for the structure in Ex. 5.2 with the following exceptions: the structure is heated ($C_t = 1.0$), the roof slope is 3 on 12 (14.0 degrees), the roof surface is slippery, and the roof is unventilated with $R \geq 30$ °F h ft^2/BTU.

Solution

As before, $C_e = 1.0$, $I = 1.2$, and $p_g = 20$ psf, but now $C_t = 1.0$ and the roof slope factor from Figure 7-2(a) is as follows:

$$C_s = 1.0 - \frac{14° - 5°}{65°} = 0.86$$

The flat roof load becomes

$$p_f = 0.7 \, C_e \, C_t \, I \, p_g$$
$$= 0.7(1.0)(1.0)(1.2)(20)$$
$$= 16.8; \text{ round to } 17 \text{ psf}$$

The balanced snow load for use with the partial loading (the sloped roof snow load by Eq. (7-2)) is

$$p_s = C_s \, p_f$$
$$= 0.86 \, (17 \text{ psf})$$
$$= 14.6; \text{ round to } 15 \text{ psf}$$

For a 5-ft purlin spacing, the full- and half-balanced loads are

full-balanced = 15 psf × 5.0 ft = 75 plf

half-balanced = ½(75 plf) = 37.5; round to 38 plf

The distribution of these partial loads along the purlins would be the same as in Ex. 5.2, as shown in **Figure V-9,** with the exception that the full- and half-balanced values are 75 and 38 plf as opposed to the 93 and 46 plf.

The example above provides the four load cases for partial loading; however, the purlins also need to be checked for two additional load cases. The first is a uniform load of 75 plf corresponding to a balanced load case. The second is another uniform load case corresponding to gable roof drifting loads, which is illustrated in Ex. 6.1.

Note that a minimum roof loading (Section 7.3.4) does not apply to this structure since we have a reasonably large roof slope [i.e., 3 on 12 = 14° is larger than $(70/W) + 0.5 = 3.3°$]. Similarly, a rain-on-snow surcharge does not apply because the roof slope of 3 on 12 is greater than the ½-on-12 slope criterion. However, even if they did apply to the structure, they would not affect the partial loads, which are based on p_s.

5.6 Example 5.4: Partial Snow Load, Continuous Purlins with Low Ground Load

Same as Ex. 5.3, except p_g = 15 psf.

Solution	In Ex. 5.3, p_g was 20 psf and the full- and half-balanced loads were 75 and 38 plf, respectively. Since the new ground snow load is 75% of that value (15/20 = 0.75), the full- and half-balanced loads are 56 and 29 plf, respectively (75% of the Ex. 5.3 values). Note that because the ground snow load is less than 20 psf, a separate *uniform* load case corresponding to the minimum roof load provisions would need to be checked.

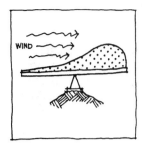

6
Unbalanced Loads

Unlike the partial loads discussed in Chapter 5 of this guide, the unbalanced loads in Section 7.6 of ASCE 7-02 are actual drift loads in most cases. Except for sawtooth-type roofs, the windward portion of the roof is the snow source and the leeward portion accumulates a percentage of the drifted snow. This is termed an "unbalanced" condition because the leeward portion has more snow than the windward portion.

The unbalanced snow loading on hip and gable roofs (Section 7.6.1), curved roofs (Section 7.6.2), sawtooth roofs (Section 7.6.3), and dome roofs (Section 7.6.4) is discussed in the following sections.

6.1 Hip and Gable Roofs

Unbalanced or across-the-ridge drift loads must be considered for the following roof slopes (in degrees):

$$\frac{70}{W} + 0.5 \leq slope < 70° \qquad \textbf{(Eq. VI-1)}$$

where W is the horizontal eave-to-ridge distance in feet. The lower bound, in essence, waives the unbalanced load requirement for narrow, low-sloped roofs. Significant drifts have not been observed for these cases, and they are not expected. Note that for small to moderate roof slopes, the horizontal eave-to-ridge distance multiplied by the roof slope is about equal to the vertical distance between eave and ridge. This elevation difference, in turn, is a measure of the space available for drift accumulation, as shown in **Figure VI-1.** Hence, designers do not need to consider across-the-ridge drifting when the space available for drift accumulation is small. In **Table VI-1,** W is

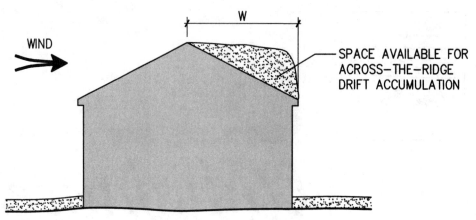

WIND

W

SPACE AVAILABLE FOR
ACROSS—THE—RIDGE
DRIFT ACCUMULATION

Figure VI-1   Space Available for Gable Roof Drift Accumulation

Table VI-1   Approximate Eave-to-Ridge Distances below which Unbalanced Loading on Hip and Gable Roofs Need Not be Considered

Roof Slope	θ (degrees)	W (ft)
1/4 on 12	1.19	101
3/8 on 12	1.79	54
1/2 on 12	2.39	37
3/4 on 12	3.58	23
1 on 12	4.76	16
3 on 12	14.0	5
4 on 12	18.4	4

the horizontal eave-to-ridge distance for various roof slopes, below which across-the-ridge drifts need not be considered.

The upper bound roof slope of 70 degrees is consistent with the roof slope factors in Figure 7-2 in ASCE 7-02. Regardless of the roof surface, $C_s = 0$ for roof slopes of 70 degrees or more for both warm ($C_t \leq 1.0$) and cold ($C_t \geq 1.1$) roofs. Because snow is presumed not to remain on such steeply sloped roofs, consideration of across-the-ridge drift loading is unwarranted.

Two unbalanced load distributions are prescribed for hip and gable roofs. For eave-to-ridge distances less than or equal to 20 ft ($W \leq 20$ ft), one distribution applies, and for eave-to-ridge distances greater than 20 ft ($W > 20$ ft), another distribution applies. The first distribution (no loading on the windward side and a uniform load of 1.5 p_s/C_e on the leeward side) applies to small- and moderate-sized roofs, specifically roofs with an eave-to-ridge distance $W \leq 20$ ft. This simple and straightforward loading is intended for use with single-family residences, the vast majority of which have eave-to-ridge distances of 20 ft or less.

Table VI-2 Unbalanced Load on Leeward Side of Nonslippery Gable Roof
with $W \leq 20$ ft and $I = 1.0$

	Thermal Factor	
Roof Slope	$C_t = 1.0$	$C_t = 1.1$
6 on 12 and less	$1.05\, p_g$	$1.16\, p_g$
8 on 12	$0.95\, p_g$	$1.16\, p_g$
12 on 12	$0.66\, p_g$	$0.89\, p_g$

The leeward drift load for these moderate-sized structures is 1.05 $C_t C_s I p_g$. This load is presented in **Table VI-2** for various roof slopes. The roof surface is assumed to be nonslippery (e.g., wood and asphalt shingles), and the structure's importance factor is assumed to be 1.0.

For wide gable roofs (i.e., $W > 20$ ft), the unbalanced load is more complex. The uniform load on the windward side is $0.3\, p_s$ and the leeward load is a uniform $1.2\,(1 + \beta/2)\,p_s/C_e$ as presented in Figure 7-5. The gable roof drift parameter (β) is defined in Eq. (7-3).

$$
\beta = \begin{cases} 1.0 & p_g \leq 20\ \text{lb/ft}^2 \\ 1.5 - 0.025 p_g & 20 < p_g < 40\ \text{lb/ft}^2 \\ 0.5 & p_g \geq 40\ \text{lb/ft}^2 \end{cases} \qquad \text{(Eq. 7-3)}
$$

As with single-family residence type gable roofs (i.e., $W \leq 20$ ft), the leeward unbalanced load is proportional to the sloped roof snow load, p_s, divided by the exposure factor, C_e. The use of p_s is consistent with the upper bound roof slope of 70 degrees in that the unbalanced loading becomes smaller and smaller as the slope approaches 70 degrees. The sloped roof load is divided by C_e to properly account for the effects of wind. In general, one expects less snow atop a fully exposed roof than a sheltered roof. On the other hand, one expects more drifting on a fully exposed roof than on a sheltered roof. Thus, the two wind effects tend to counteract each other. This is generally consistent with the unbalanced load being proportional to p_s/C_e, which results in a leeward unbalanced load that is independent of C_e.

As noted in the Commentary (Section C7.6.1), the gable roof drift parameter, β, is proportional to the percentage of snow which is expected to drift across the ridge line. It is a decreasing function of the ground snow load, p_g, as shown in Eq. (7-3). As shown in Chapter 7, this is generally consistent with the interrelationship between the size of a drift and the amount of snow in the source area; that is, as the source area increases, a lower *percentage* of the source area snow eventually accumulates in the drift.

Table VI-3 presents the unbalanced load on the leeward side of a wide gable roof (i.e., $W > 20$ ft) for various values of the ground snow load and the thermal factor. The roof slope is 6 on 12 or less, the roof surface is assumed to be nonslippery, and the importance factor, I, is assumed to be 1.0.

Table VI-3   Unbalanced Load on Leeward Side of Nonslippery Gable Roof with $W > 20$ ft, Roof Slope $\leq$ 6 on 12, and $I = 1.0$

	Thermal Factor	
Ground Snow Load	$C_t = 1.0$	$C_t = 1.1$
20 psf and less	$1.26\, p_g$	$1.39\, p_g$
30 psf	$1.16\, p_g$	$1.28\, p_g$
40 psf and above	$1.05\, p_g$	$1.16\, p_g$

A comparison of **Table VI-2** (narrow roofs) and **Table VI-3** (wide roofs) indicates that the code-specified leeward unbalanced load on a wide ($W > 20$ ft) roof is larger than or at least equal to the corresponding value for a narrow roof ($W \leq 20$ ft). This trend is consistent with a database of gable roof case histories developed by O'Rourke and Auren (1997). In this database of large drifts (many from insurance company files, most involving structural damage), only four of the 28 buildings had $W < 20$ ft. Loss experience suggests that gable roof drifting is more of a problem for wide roofs, and hence, arguably, the unbalanced loading for this class of roof ($W > 20$ ft) should be larger.

6.2   Example 6.1: Unbalanced Snow Load, Narrow Gable Roof

Determine the unbalanced loading for a single-family residence with a symmetric 6-on-12 gable roof and an eave-to-eave distance of 38 ft. The home has a well-insulated "cold roof" ($C_t = 1.1$) with asphalt shingles (i.e., nonslippery roof surface) and is located in an area where $p_g = 25$ psf.

Solution

Since the roof is symmetric, the eave-to-ridge distance is $38/2 = 19$ ft; hence, the simple unbalanced provision (i.e., $W \leq 20$ ft) applies. As previously noted, the exposure coefficient, C_e, is not needed since it cancels out. From **Table VI-2**, a leeward load for a 6-on-12 slope and $C_t = 1.1$ equals $1.16\, p_g$. Hence, the unbalanced load is $1.16(25 \text{ psf}) = 29$ psf, as shown in **Figure VI-2**.

6.3   Example 6.2: Unbalanced Snow Load, Wide Gable Roof

Determine the unbalanced snow loading for the structure in Ex. 4.3.

Solution

Since the eave-to-ridge distance is larger than 37 ft for this ½-on-12 roof (see **Table VI-1**), unbalanced loads need to be considered. From Ex. 4.3, it has been determined that $p_s = 25$ psf, $C_e = 1.0$, and $p_g = 30$ psf. For a ground snow load of 30 psf, the gable roof drift parameter, β, becomes

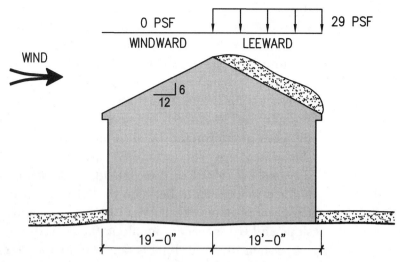

Figure VI-2 Unbalanced Snow Load and Elevation for Ex. 6.1

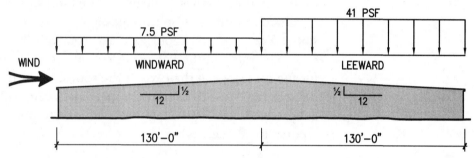

Figure VI-3 Unbalanced Snow Load and Elevation for Ex. 6.2

$$\beta = 1.5 - 0.025 p_g$$
$$= 1.5 - 0.025(30 \text{ psf})$$
$$= 0.75$$

Since $W > 20$ ft, the windward side load is $0.3\, p_s = 0.3(25) = 7.5$ psf, while the leeward load is $1.2(1 + \beta/2)p_s/C_e = 1.2(1 + .75/2)25 \text{ psf}/1.0 = 41$ psf. The resulting unbalanced load is shown in **Figure VI-3.** Any partial loading (e.g., continuous purlins) would be a separate load case.

6.4 Curved Roofs

Because of the complicated geometry for a curved roof, the Standard presents the balanced and unbalanced snow load cases in Figure 7-3.

6.4.1 Balanced Loads

For the case where the slope at the eave of the curved roof is less than 30 degrees (i.e., Case 1), the balanced load is uniform near the crown and is trapezoidal adjacent to the eaves. The uniform load extends over the low-sloped portion of the roof where $C_s = 1.0$ as determined from Figure 7-2. For example, if $C_t = 1.1$ and the roof had an unobstructed, slippery surface, the uniform load region would extend to the point where the roof slope was greater than 10 degrees (see Figure 7-2b). Beyond the uniform load region, the balanced load is assumed to decrease linearly to a value corresponding to the sloped roof snow load at the eave. The roof snow load at the eave is determined by multiplying the flat roof snow load by the roof slope factor associated with the slope at the eave. To simplify the snow load diagram, the sloped roof snow load, p_s, is established at selected points and a linear interpolation between the selected locations is used. This method closely approximates a curved load diagram and avoids evaluating the roof slope and C_s for every point along the curved roof surface.

When the eave slope is between 30 and 70 degrees (i.e., Case 2), the balanced load is uniform near the crown (where $C_s = 1.0$) and there are two trapezoidal loads—a middle trapezoid and an edge trapezoid—located near the eave. The middle trapezoid starts where $C_s < 1.0$ and terminates where the roof slope is 30 degrees. The intensity of the balanced load at the termination point is p_f multiplied by C_s corresponding to a 30-degree roof slope. The edge trapezoid extends from the 30-degree roof slope to the eave. The value of the balanced load at the eave equals p_f multiplied by C_s for the roof slope at the eave. If the roof slope at the eave is greater than 70 degrees (i.e., Case 3) then the trapezoidal edge load is replaced with a triangular edge load. The triangular edge load terminates where the roof slope is equal to 70 degrees.

6.4.2 Unbalanced Loads

The unbalanced load case for a curved roof has zero load on the windward side and, for the simplest case, a trapezoidal load on the leeward side. In all cases, the unbalanced load at the crown is $0.5\ p_f$. For Case 1, where the roof slope at the eave is less than 30 degrees, the leeward unbalanced load increases from the $0.5\ p_f$ value at the crown to a value of $2p_f C_s /C_e$ at the eave. As with the balanced load case, C_s is the roof slope factor corresponding to the eave roof slope. In a sense, the value at the eave is a multiple of the sloped roof load divided by the exposure factor (p_s/C_e), similar to the formulation for unbalanced loads on gable roofs.

The trapezoidal load is composed of a uniform load, equal to $0.5\ p_f$, with a triangular surcharge. The triangular surcharge is maximum at the eave. This is generally consistent with the curved roof drift shown in **Figure VI-4.** That is, moving from crown to eave, the vertical distance from the roof surface to the crown elevation gets larger and consequently there is more space for drift accumulation.

Again for Case 1 (eave slope < 30 degrees or roughly 7 on 12), the average leeward unbalanced load for a curved roof with a nonslippery surface, $C_t = 1.0$ and $I = 1.0$, would be $0.88p_g$ for $C_t = 1.0$ and $0.96p_g$ for $C_t = 1.1$. A comparison of corresponding values in **Table VI-2** (roof slope ≤ 6 on 12 < 30 degrees) and **Table VI-3** ($p_g \geq 40$ psf) indicates that the curved roof aver-

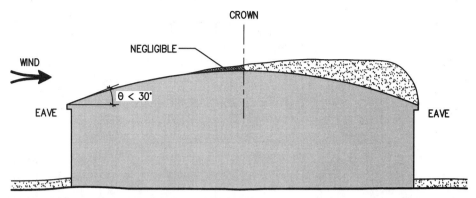

Figure VI-4 Typical Curved Roof Drift Formation

age unbalanced load is less than the unbalanced load for gable roofs. However, the curved roof distribution is arguably more realistic.

The unbalanced loading for Cases 2 and 3 (eave slopes $\geq$ 30 degrees) mimics that outlined for balanced loading. There is a linear variation from $0.5 p_f$ at the crown to $2 p_f C_s / C_e$ at the eave or the 70-degree point. Again, the C_s factors used are those for the roof slope at the point of interest, such as the 30-degree point, 70-degree point, or roof slope at the eave.

Like the gable roof unbalanced load, the curved roof unbalanced load applies to a specific range of roof slopes. However, unlike all the previous provisions, which were based on the tangent slope, the limits for curved roofs are based on the secant slope. Specifically, unbalanced loads are not considered if the slope of a straight line from the eave (or 70-degree point) to the crown is less than 10 degrees or greater than 60 degrees.

6.5   Example 6.3: Balanced and Unbalanced Snow Load, Curved Roof

Determine the balanced and unbalanced load for a 450-seat heated theater at a windswept suburban location with p_g = 25 psf. The roof is a circular arc with a rise of 15 ft and a span of 80 ft, as shown in **Figure VI-5.** The built-up roof has an aggregate surface, and the roof system is unventilated.

Solution

Balanced Load: From Table 7-2, C_e = 0.9 for a fully exposed roof in Terrain Category B. For a heated unventilated roof, C_t = 1.0 from Table 7-3. Since more than 300 people congregate in one area, the facility is a Category III structure as described in Table 1-1 and hence I = 1.1 from Table 7-4. Therefore,

$$p_f = 0.7 C_e C_t I p_g$$
$$= 0.7(0.9)(1.0)(1.1)(25 \text{ psf})$$
$$= 17 \text{ psf}$$

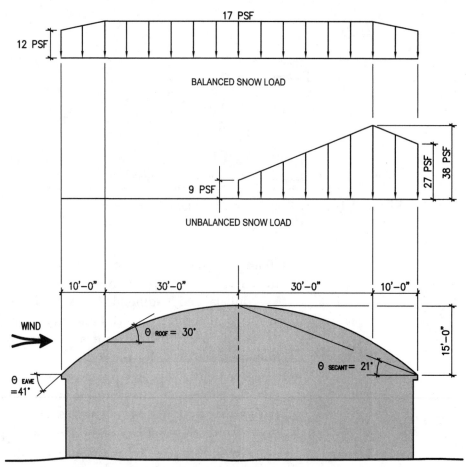

Figure VI-5   Balanced and Unbalanced Snow Loads for Curved Roof in Ex. 6.3

From **Figure VI-5,** the slope of the eave is 41 degrees; hence, Case 2 of Figure 7-3 applies.

The thermal factor, C_t, is 1.0 and the aggregate roof surface is not slippery; therefore, the solid line in Figure 7-2a is used to determine C_s. From Figure 7-2a, $C_s = 1.0$ for roof slopes ≤ 30 degrees; therefore, $C_s = 1.0$ for all areas where the tangent roof slope is ≤ 30 degrees. Using the formulas in the Commentary, the C_s factor for the tangent slope at the eave equals

$$C_s^* = 1.0 - \frac{41° - 30°}{40°} = 0.73$$

Hence, the balanced load from the crown to the 30-degree point is equal to 17 psf while the value at the eave is $p_f C_s^* = 17 \text{ psf}(.73) = 12$ psf. This balanced load is shown in **Figure VI-5.**

Unbalanced Load: Since the secant angle from crown to eave is greater than 10 degrees, unbalanced loads must be considered. For Case 2, the load at the crown is $0.5\, p_f = 0.5(17 \text{ psf}) = 9$ psf, at the 30-degree point is $2p_f C_s^*/$

$C_e = 2(17 \text{ psf})(1.0)/0.9 = 38$ psf, and at the eave is $2p_f C_s^*/C_e = 2(17$ psf$)(0.73)/0.9 = 27$ psf. The resulting unbalanced load is also sketched in **Figure VI-5.**

Since the vertical angle from the eaves to the crown (secant angle) was greater than 10 degrees, unbalanced loads were considered. Correspondingly, a minimum roof load, as prescribed in Section 7.3.4, need not be considered.

For curved roofs with an eave slope greater than 30 degrees (Case 2 or 3 in Figure 7-3), the unbalanced load begins to decrease downwind of the 30-degree tangent point. This reduction is related to steep roof effects including possible sliding near the eaves. However, if an adjacent roof is closeby (specifically within 3 ft), the potential for steep roof effects is significantly decreased and the reduction is not allowed. For example, if there were an abutting roof in Ex. 6.3, a uniform load (= 38 psf) would be distributed from the 30-degree point to the eave.

6.6 Sawtooth-Type Roofs

Unbalanced loads on multiple folded-plate sawtooth and multiple barrel vault roofs are presented in Section 7.6.3. The prescribed loading is similar to that for a curved roof, i.e., $0.5\ p_f$ at the high point and $2\ p_f/C_e$ at the low point. However, unbalanced loads on sawtooth roofs are fundamentally different from unbalanced loads for gable, curved, or even dome roofs. For all of the previous roof types, there are windward and leeward sides, and the unbalanced load was due to drifting. The authors envision unbalanced loading on sawtooth roofs as arising from snow sliding from the high point to the low point. Note, however, that the average unbalanced load for a sawtooth roof is larger than the balanced load, p_f. This is inconsistent with the view that the unbalanced load is due solely to sliding or redistribution of the flat roof load. Other notable differences are: (1) the unbalanced load for a curved roof is a function of C_s, whereas $C_s = 1.0$ for the sawtooth-type roofs as per Section 7.4.4; (2) for curved roofs, unbalanced loads are not considered for secant slopes of less than 10 degrees or greater than 60 degrees, while for sawtooth-type roofs the low roof slope cutoff is 3/8 on 12 (1.79 degrees) and there is no high roof slope limit; and (3) for sawtooth-type roofs, a limitation on the unbalanced load at the low point is prescribed, based on space available for the snow accumulation.

6.7 Example 6.4: Unbalanced Snow Load, Sawtooth Roof

Determine the unbalanced load for a retail greenhouse roof shown in **Figure VI-6**. The R-value for the roof is 1.5 °F h ft^2/BTU and the facility is continuously heated. It is located adjacent to taller establishments in a suburban shopping mall where $p_g = 35$ psf.

Solution	**Balanced Load:** Since the roof is neither fully exposed nor completely sheltered, it is classified as partially exposed. Hence, for its suburban location (Terrain Category B), $C_e = 1.0$ from Table 7-2. Since it is continuously heated with a roof R-value < 2 °F h ft^2/BTU, the thermal factor $C_t = 0.85$ from Table 7-3. Since the retail greenhouse is considered to be more like a store than an agricultural facility, the importance factor, I, is 1.0. Hence,

$$p_f = 0.7 C_e C_t I p_g$$
$$= 0.7(1.0)(0.85)(1.0)(35 \text{ psf})$$
$$= 21 \text{ psf}$$

Unbalanced Loads: Since the roof slope is larger than 3/8 on 12, unbalanced loads need to be considered. The prescribed load at the high point is $0.5\,p_f = 0.5(21 \text{ psf}) = 11$ psf. At the low point, the prescribed load is $2p_f/C_e = 2(21 \text{ psf})/1.0 = 42$ psf. However, the load at the low point is limited by the space available. Given a 35-psf ground snow load site, the snow density $\gamma = 0.13\,p_g + 14 = 0.13(35 \text{ psf}) + 14 = 19$ pcf. Hence, the maximum snow load cannot be greater than the load at the high point (11 psf) plus 4 ft of snow at 19 pcf or 4 ft (19 pcf) = 76 psf. Since the unbalanced load of 42 psf is less than 11 psf + 76 psf (= 88 psf maximum permissible load), the load at the low point is not truncated and it remains 42 psf, as shown in **Figure VI-6**.

6.8 Domes

Balanced roof snow loads for domes are the same as the corresponding curved roof balanced loads. Unbalanced loads on domes are based on the corresponding curved roof unbalanced load applied on a 90-degree downwind sector. To each side of the 90-degree sector, there is a 22.5-degree sector where the unbalanced load decreases linearly to zero. The unbalanced load on the remaining 225-degree upwind sector is zero.

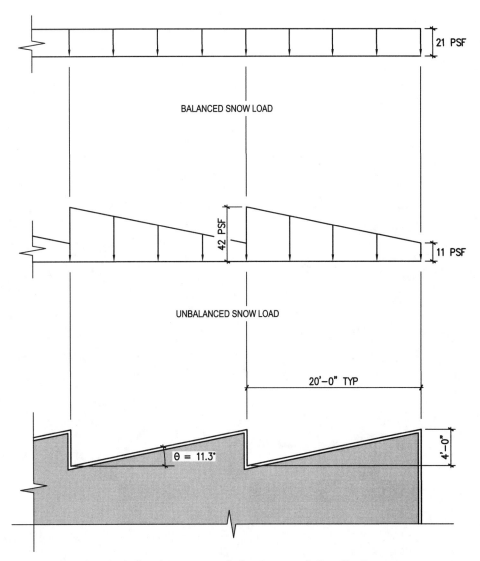

21 PSF

BALANCED SNOW LOAD

42 PSF

11 PSF

UNBALANCED SNOW LOAD

20'–0" TYP

4'–0"

θ = 11.3°

Figure VI-6   Balanced and Unbalanced Snow Loads for Sawtooth Roof in Ex. 6.4

7

Drifts on Lower Roofs

In the past, roof step drifting accounted for roughly 75% of all snow-related roof damage (O'Rourke et al. 1982). With the advent of modern code provisions, the amount of snow-related roof damage caused by roof step drifting has decreased. Roof step drifts now account for approximately 25% to 50% of all snow-related losses.

Drifts accumulate differently when the higher level roof is on the windward side than they do when the higher level roof is on the leeward side. As shown in **Figure VII-1,** leeward drifts are nominally triangular in shape, while windward drifts are more complex. Windward drifts often start out as quadrilateral shapes because a vortex forms when wind impinges on the vertical wall directly beyond the drift. However, as the windward roof step drift grows in height and the vertical distance between the top of the drift and the top of the wall diminishes, the wind is redirected over the top of the wall as opposed to impinging on the wall. At this point the windward drift begins to morph into a nominally triangular shape. As shown in **Figure VII-2,** when the wind streamlines no longer hit the wall, the shaded area fills with snow, resulting in a triangular shape.

The ASCE 7-02 provisions relate the size of a roof step drift to the amount of snow in the source area upwind of the step. The amount of snow is related to the ground snow load, p_g, and the upwind fetch area or length. For the windward drift atop Roof A in **Figure VII-1,** the fetch is the lower roof length upwind of the windward step, ℓ_ℓ. Similarly for the leeward step at Roof C, the fetch is the upper roof length, ℓ_u. In cases where the elevation difference at the windward step (i.e., elevation difference between Roof A and Roof B) is small, the leeward drift fetch will approach the sum of the upper and *lower* level roof lengths upwind of the step ($\ell_\ell + \ell_u$).

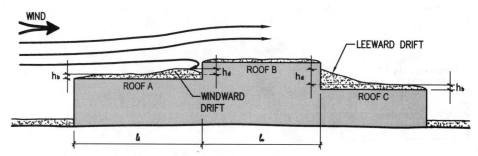

Figure VII-1   Windward and Leeward Snow Drifts

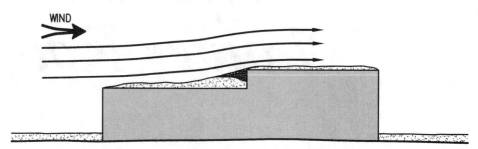

Figure VII-2   Windward Drift Morphing from Quadrilateral to Triangular Shape

Hence, if the windward roof step is easily filled, then both upper and lower upwind roofs serve as the source area for the leeward drift. In such cases, it is conservative to use the sum of the upwind roof lengths as the leeward drift fetch. A more exact approach for roofs with two potential upwind snow sources is presented in Chapter 12.

7.1   Leeward Drift

The roof step relations are empirical, as they are based on an analysis of case histories. For example, the leeward relation is based on an analysis of approximately 350 nominally triangular drifts from insurance company files and other sources (O'Rourke et al. 1985, 1986). Multiple regression analyses suggested the following relationship between the surcharge drift height, h_d, defined as the drift height above the balanced snow, the upwind fetch, ℓ_u, and the observed ground snow load, p'_g, for leeward drifts.

$$h_d = 0.61\sqrt[3]{\ell_u}\sqrt[4]{p'_g + 10} - 2.2$$

(Eq. VII-1)

The relative accuracy of the relation in **Eq. (VII-1)** is shown in **Figure VII-3** wherein observed surcharge heights are plotted versus the predicted drift surcharge height by the regression equation. Note that most of the observed data points fall within a factor of two of the predicted value.

The ground snow load, p'_g, in **Eq. (VII-1)** is the observed case history value, not the 50-yr mean recurrence interval (MRI) value for the site. The

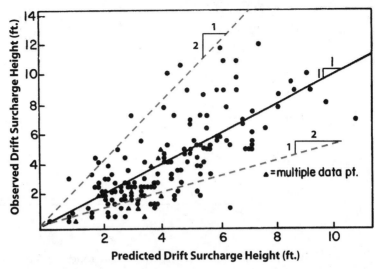

Figure VII-3 Observed Drift Surcharge Height versus Predicted Drift Surcharge Height, per Eq. (VII-1)

Source: O'Rourke et al. 1986.

observed ground snow load is actually less than half the 50-yr value for a majority of the case histories. Although the observed ground snow load was typically less than the 50-yr MRI, the case history database arguably represented appropriate design drifts because more than 40% of the case histories involved structural failure of one kind or another. However, the ASCE 7 Snow Task Committee wanted an equation that used the 50-yr ground snow load because the 50-yr value is already being used in ASCE 7. To utilize the 50-yr value for p_g and to predict reasonable drift heights that were close to those observed in the case histories, the whole relation in **Eq. (VII-1)** was multiplied by a modification factor, α, which is less than one. Hence, the relation for the surcharge drift height became

$$h_d = \alpha\left[0.61\sqrt[3]{\ell_u}\sqrt[4]{p_g + 10} - 2.2\right]$$ (Eq. VII-2)

where p_g is the 50-yr ground snow load for the site per ASCE 7.

Table VII-1 shows the effect of various values for the modification factor, α. For a modification factor of 0.5, 55% of the observed drifts were larger than the values predicted by **Eq. (VII-2)**. On the other hand, for a modification factor of 0.9, only 21% of the observed drift exceeded the predicted values from **Eq. (VII-2)**. Based on engineering judgment, the ASCE 7 Snow Task Committee chose a modification factor of 0.7. As such, the predicted drift exceeded the observed drift for about two-thirds of the case histories. Using a reduction factor of 0.7, the relation for the surcharge drift height becomes

$$h_d = 0.43\sqrt[3]{\ell_u}\sqrt[4]{p_g + 10} - 1.5$$ (Eq. VII-3)

Table VII-1 Effect of Modifying Factor on Eq. (VII-2)

Modifying Factor α	Percentage of Case Histories with Observed Drift > Predicted Drift
1.0	17
0.9	21
0.8	28
0.7	32
0.6	41
0.5	55

where p_g is the 50-yr ground snow load for the site of interest.

Figure 7-9 in ASCE 7-02 is a plot of **Eq. (VII-3).** The width of the drift is prescribed to be four times the surcharge height (i.e., $w = 4\ h_d$) as long as the drift does not become "full." The assumed rise-to-run of 1:4 is based on an analysis of 101 case histories for which both the surcharge drift height and the width of the drift were available. **Figure VII-4** shows a scattergram of the drift height versus drift width data. Considering all the data points, the slope of the regression line is 0.227 (a rise-to-run of 1:4.4). However, when the "full" drifts (drifts that have a total height within 6 in. of the roof elevation) and non-full drifts are separated, the full drifts had a rise-to-run of about 1:5 and the non-full drifts had a slope of about 1:4. This suggests that the drifts initially form with a rise-to-run of about 1:4, and when the drift becomes full, additional snow accumulates at the toe of the drift, resulting in a flatter slope. Hence, as prescribed in Section 7.7.1, if the drift is full (i.e., $h_d = h_c$, where h_c is the space above the balanced snow available for drift formation), then the drift width, w, becomes $4\ h_d^2 / h_c$ with a maximum of $8\ h_c$. The full-drift relation for w was determined by equating the cross-sectional area of a height limited triangular drift (i.e., $0.5\ h_c w$) to the cross-sectional area of a height unlimited drift with the same upwind fetch and ground load (i.e., $0.5\ h_d(4\ h_d)$). The upper limit of $8\ h_c$ for the width of a full drift is based on the concept of an aerodynamically streamlined drift (rise-to-run of approximately 1:8) for which significant additional accumulation is not expected.

Eq. (VII-3) provides the surcharge height of the design drift for leeward wind. To convert height to an equivalent snow load, the density or unit weight of the snow is required. ASCE 7-02 uses the following relationship for the unit weight of snow, γ, in pounds per cubic foot (pcf):

$$\gamma = 0.13 p_g + 14 \le 30 \text{ pcf} \tag{Eq. 7-4}$$

where the ground snow load, p_g, has units of pounds per square foot (psf). This relation was originally developed by Speck (1984). Eq. (7-4) illustrates

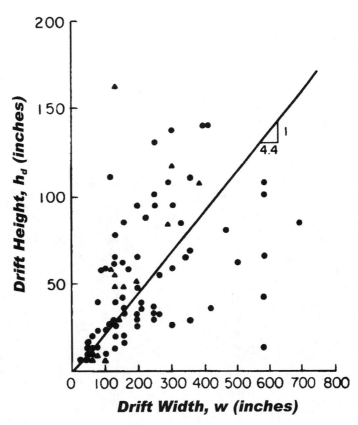

Figure VII-4   Surcharge Drift Height versus Drift Width
Source: O'Rourke et al. 1985.

that the snow density is an increasing function of snow depth. **Figure VII-5** is a plot of snowpack density (pcf) versus snowpack depth (inches) as predicted by the ASCE 7-02 relation (Eq. (7-4)). Due to an upper limit of 30 pcf, the density is constant for depths greater than 49 in. or ground loads greater than 123 psf. At shallower snow depths, the formula yields roughly a 1-pcf increase in density for every 4 in. or so of additional depth.

Figure VII-6 is a plot of snow load (psf) versus snow depth (inches). It includes a density relation from Tabler (1994) for snow before the onset of melt. Notice that these two independently developed unit weight relations provide remarkably similar snow loads for snow depths less than 4 ft. Also, both curves (ASCE 7-02 and Tabler) are convex (i.e., the density or unit weight is an increasing function of depth). This increase is due, at least in part, to self-compaction due to the weight of the overburden snow.

The Tobiasson and Greatorex (1996) relation between 50-yr load and 50-yr depth from **Eq. (II-1)** and **Figure II-1** is shown as a dashed line in **Figure VII-6.** The Tobiasson and Greatorex relation suggests lower loads for the same depth of snow. The differences are due in large part to the nature of the two sets of relations. The ASCE 7-02 and Tabler relations are based on simultaneous measurements of load and depth. On the other hand, the

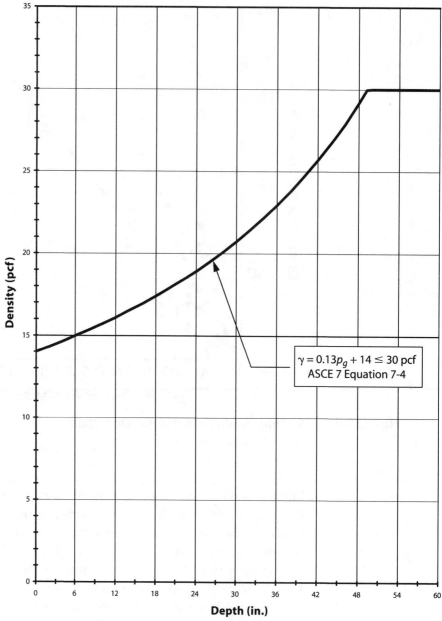

Figure VII-5 Snowpack Density versus Snowpack Depth, per Eq. (7-4)

Tobiasson and Greatorex formula relates a maximum annual snow depth to a maximum annual snow load (50-yr ground snow depth to 50-yr ground snow load). For a common scenario when the maximum depth occurs earlier in the winter than the maximum load, the Tobiasson and Greatorex conversion density for this maximum depth would be less than the actual density when the load reached maximum.

Although the two sets of density relations provide different answers, both are arguably appropriate for their intended purposes. ASCE 7-02 (Eq. (7-4)) and Tabler convert a snow depth at a point in time into a snow load

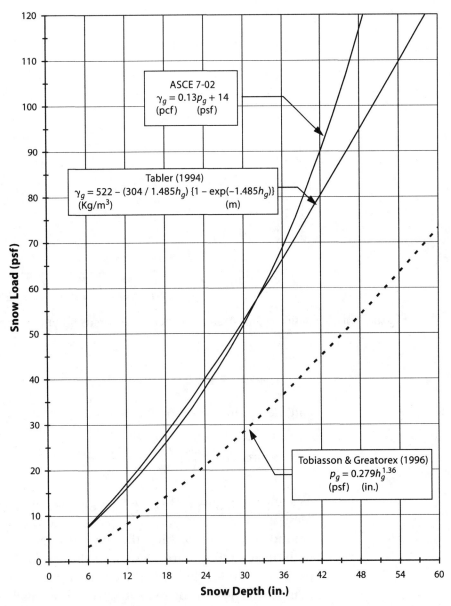

Figure VII-6   Snow Load versus Snow Depth

at the same point in time. Tobiasson and Greatorex (**Eq. (II-1)**) relate a 50-yr snow depth at a point in time to a 50-yr snow load, possibly at another point in time.

Eqs. (VII-1) through (**VII-3**) indicate that the drift size is an increasing function of both the ground snow load and the upwind fetch. In other words, the bigger the snow source, the bigger the drift. However, the increase is not linear. For example, doubling either the upwind fetch or the ground snow load results in less than a doubling of the drift size. This is illustrated in **Figure VII-7,** which is a plot of the ratio of the cross-sectional area of the drift to the upwind snow source area versus the 50-yr ground snow load where the drift area is

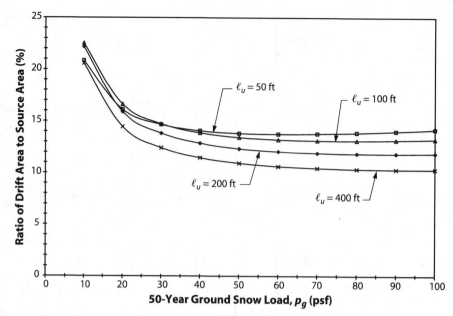

Figure VII-7   Ratio of Drift Area to Source Area versus 50-yr Ground Snow Load

$$\text{Drift Area} = \tfrac{1}{2} h_d w = 2\, h_d^2 \qquad\qquad \textbf{(Eq. VII-4)}$$

and the upwind snow source area is

$$\text{Source Area} = \ell_u \frac{p_g}{\gamma} = \frac{\ell_u p_g}{0.13 p_g + 14} \qquad\qquad \textbf{(Eq. VII-5)}$$

As shown in **Figure VII-7,** the "design" leeward drift is 10% to 25% of the "design" snow source area. The percentage is a decreasing function of the ground snow load, p_g, and the upwind fetch, ℓ_u, although less so for ℓ_u. Both of these trends are sensible. If the upwind fetch is small or the snow-pack depth is shallow, then a typical wind event could easily remove or transport almost all of the snow from the small source area. Hence, it is likely that a significant fraction of a small snow source area could end up in the drift. Conversely, for larger fetch areas and/or deep snowpacks, a smaller percentage of snow is transported. Note that the range of percentages (10% to 25%) in **Figure VII-7** is based on the 50-yr ground snow load, as are those in ASCE 7-02 (see **Eq. (VII-3)** and Figure 7-9). When the ratio of drift area to source area is compared with observed ground snow loads from case studies (**Eq. (VII-1)**) instead of the 50-yr load, the percentages double to roughly 20% to 50%. This occurs because the 0.7 modification factor used in ASCE 7-02 is applied to both the surcharge height, h_d, and the width ($w = 4h_d$) for a given source area. In other words, 20% and 50% of the upwind snow source typically ended up in the case history drifts, while for our code relations, in which the snow is characterized by the 50-yr value, the "design" drift is about 10% to 25% of the "design" upwind source area.

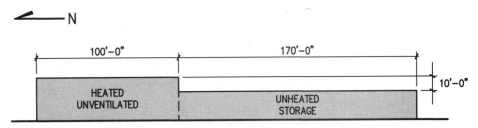

Figure VII-8 West Elevation of Stepped Roof Structure for Ex. 7.1

7.2 Windward Drift

Eq. (VII-3) and Figure 7-9 can be used to determine the windward drift height as well, with some modifications. In **Eq. (VII-3)** and Figure 7-9, ℓ_u is replaced with ℓ_ℓ and then the calculated height is multiplied by 0.75. Case histories suggest that windward steps trap snow less efficiently than leeward steps, resulting in a reduced drift height. More detailed justification for the three-quarters factor is provided in Chapter 8 of this guide. In all cases, the triangular drift surcharge is superimposed on the sloped roof load for the lower roof.

7.3 Example 7.1: Roof Step Drift Load

Determine the design snow loads for the structure in **Figure VII-8.** This ground snow load, p_g, is 40 psf, the heated portion is of ordinary importance, and the site is in flat open country (Terrain Category C) with no trees or nearby structures offering shelter. Both roofs have ¼-on-12 slopes in the east-west direction to internal drains.

Solution

Balanced Load (Upper Roof Level): Because the building is located in Terrain Category C and the upper roof is fully exposed, $C_e = 0.9$ from Table 7-2. For a heated space with an unventilated roof, the thermal factor, C_t, equals 1.0 from Table 7-3 and the importance factor, I, equals 1.0 from Table 7-4. Hence, the upper flat roof snow load is

$$p_f = 0.7 C_e C_t I p_g$$
$$= 0.7(0.9)(1.0)(1.0)(40 \text{ psf})$$
$$= 25 \text{ psf}$$

For a roof slope of ¼-on-12, $C_s = 1.0$ irrespective of roof material/surface. Hence, the balanced sloped roof snow load for the upper roof is also 25 psf.

Balanced Load (Lower Roof Level): As stated in the problem, the site is considered Terrain Category C. The lower roof, however, is sheltered by the presence of the upper level roof. Therefore, the lower roof is classified as partially exposed and $C_e = 1.0$ from Table 7-2. For an unheated space, the thermal factor, C_t, equals 1.2 from Table 7-3. Although this is a storage space, it is not considered prudent to classify this building as a "minor storage facility," as described in Category I in Table 1-1, because of its large footprint. Therefore, the building structure is classified as Category II in Table 1-1 with an importance factor, I, of 1.0 per Table 7-4. Hence, the balanced load on the lower level roof becomes

$$p_s = 0.7 C_e C_t C_s I p_g$$
$$= 0.7(1.0)(1.2)(1.0)(1.0)(40 \text{ psf})$$
$$= 34 \text{ psf}$$

Drift Loads: The snow density is determined from p_g using Eq. (7-4) below:

$$\gamma = 0.13 \, p_g + 14$$
$$= 0.13(40 \text{ psf}) + 14$$
$$= 19 \text{ pcf}$$

The balanced snow depth on the lower level roof is

$$h_b = \frac{p_s}{\gamma} = \frac{34 \text{ psf}}{19 \text{ pcf}} = 1.8 \text{ ft}$$

Hence, the clear height above the balanced snow is

$$h_c = 10 - h_b = 10 - 1.8 = 8.2 \text{ ft}$$

By inspection, $h_c / h_b > 0.2$; therefore, enough space is available for drift formation, and drift loads must be evaluated.

Leeward Drift: For a wind out of the north, the upwind fetch for the resulting leeward drift is the length of the upper level roof ($\ell_u = 100$ ft). Hence, the surcharge drift height is

$$h_d = 0.43 \sqrt[3]{\ell_u} \sqrt[4]{p_g + 10} - 1.5$$
$$= 0.43(100 \text{ ft})^{1/3}(40 \text{ psf} + 10)^{1/4} - 1.5$$
$$= 3.8 \text{ ft}$$

Windward Drift: For a wind out of the south, the upwind fetch for the resulting windward drift is 170 ft. Hence, the surcharge drift height is

$$h_d = 0.75 \left[0.43 \sqrt[3]{\ell_u} \sqrt[4]{p_g + 10} - 1.5 \right]$$
$$= 0.75 \left[0.43(170)^{1/3}(40 \text{ psf} + 10)^{1/4} - 1.5 \right]$$
$$= 3.6 \text{ ft}$$

Thus, the leeward drift controls, and $h_d = 3.8$ ft. Since the drift is not full ($h_c > h_d$), the drift width is four times the drift height:

$$w = 4h_d = 4(3.8 \text{ ft}) = 15 \text{ ft}$$

and the maximum *surcharge* drift load is the drift height times the snow density:

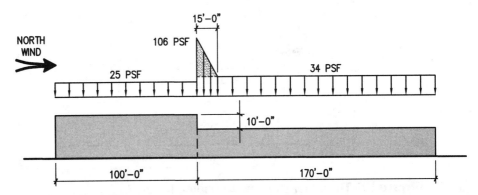

Figure VII-9 Roof Step Snow Loading for Ex. 7.1

$$p_d = h_d\gamma = 3.8 \text{ ft } (19 \text{ pcf}) = 72 \text{ psf}$$

The total load at the step is the balanced load on the lower roof plus the drift surcharge (34 + 72 = 106 psf), as shown in **Figure VII-9.**

Due to the comparatively large ground snow load ($p_g >$ 20 psf), the minimum roof load (Section 7.3.4) is 20 I or 20 psf for both the upper level and lower level roofs, and therefore does not govern. Also, due to the large ground snow load, the rain-on-snow surcharge does not apply (see Section 7.10).

Note that the windward and leeward drift heights are calculated separately, and the larger value is used to establish the design drift loading. This approach (i.e., using the *larger* of the drift heights as opposed to the *sum* of the two drift heights) is specifically mentioned in Section 7.7.1. Based on this design approach, one might assume that wind only blows from one direction throughout the winter season; however, that is not the case. In fact, it is possible to have a 180-degree shift in wind direction during a single storm event. For example, consider a storm that passes from west to east over a site. Due to the counter-clockwise rotation of the wind around the low pressure point, the site initially experiences the wind coming from the south (when the low is located to the west of the site); then, as the low pressure point moves over the site, the site experiences the wind coming from the north (when the low is located to the east).

So it is possible to have both windward and leeward contributions to the same drift formation. The approach of choosing the larger independent value for the design drift loading illustrates the empirical nature of the roof step drift provisions. That is, the leeward case history drifts, upon which the provisions are based, are due to either all leeward drifting or some combination of leeward and windward drifting. Hence, the extent to which leeward and windward drifting are both present is already reflected in the *observed* drift height. Therefore, adding the *design* leeward to the *design* windward would result in unrealistic drifts that are much larger than the observed.

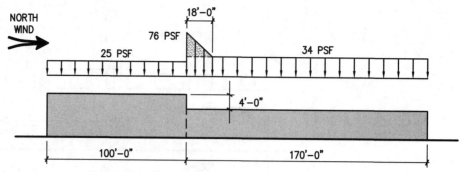

Figure VII-10   Roof Step Snow Loading for Ex. 7.2

Finally, note that in evaluating the windward drift, the full upwind fetch of 170 ft is used as opposed to some effective fetch that accounts for the space occupied by the drift itself. This is also a result of the empirical nature of the drift relations. That is, the observed drift heights were regressed against the full upwind fetch.

7.4   Example 7.2: Roof Step Drift, Limited Height

Same as Ex. 7.1, except the elevation difference at the roof step is 4 ft.

Solution

The depth of the balanced snow on the lower level roof remains 1.8 ft, but now the clear height, h_c, is 4.0 ft – 1.8 ft = 2.2 ft. Note that $h_c/h_b = 2.2$ ft/1.8 ft = 1.2 > 0.2; therefore, drift loads need to be considered. In this case, the drift surcharge height will be limited by h_c ($h_d = 3.8$ ft > h_c) and the maximum surcharge drift load is

$$p_d = h_c \gamma = 2.2 \text{ ft } (19 \text{ pcf}) = 42 \text{ psf}$$

The balanced snow load on the lower level roof remains 34 psf. Thus, the total (balanced plus drift) snow load at the roof step is 76 psf (34 psf + 42 psf, respectively). Because the drift is full, the width is increased. Recalling that the unlimited leeward drift height (i.e., $\ell_u = 100$ ft, $p_g = 40$ psf) was 3.8 ft from Ex. 7.1, the drift width equals

$$w = \frac{4(h_d)^2}{h_c} = \frac{4(3.8 \text{ ft})^2}{2.2 \text{ ft}} = 26 \text{ ft}$$

But the drift slope need not exceed a rise-to-run of 8 (aerodynamic streamlining); thus,

$$w_{\max} = 8h_c = 8(2.2 \text{ ft}) = 18 \text{ ft}$$

In this case, $8h_c$ controls and the design snow loads are shown in **Figure VII-10.**

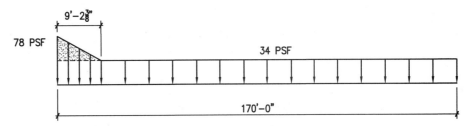

Figure VII-11 Drift Loading for Lower, Separated Roof in Ex. 7.3

7.5   Example 7.3: Roof Step Drift, Adjacent Structure

Same as Ex. 7.1, except the unheated storage facility is separated from the heated, unventilated roof facility by 8 ft.

Solution

The balanced load on the heated space remains unchanged at 25 psf. Although the heated space is no longer adjacent to the unheated space, it still serves as an obstruction (refer to the footnotes for Table 7.2). Given h_o = 10 ft and the separation distance of 8 ft < 10 h_o = 100 ft, the heated space qualifies as an obstruction for the roof of the unheated storage facility.

The unmodified leeward and windward drift heights for a roof step without a separation are still 3.8 ft and 3.6 ft, respectively. Hence, the leeward drift governs in this case. Based on the building separation of s = 8.0 ft, the modification factor for an adjacent structure is

$$\text{factor} = \frac{20 - s}{20} = \frac{20 - 8.0}{20} = 0.6 \qquad \text{(see Section 7.7.2)}$$

Therefore, the modified drift height is 0.6(3.8 ft) = 2.3 ft. The maximum surcharge load is 2.3 ft (19 pcf) = 44 psf, and the drift width is

$$w = 4(2.3 \text{ ft}) = 9.2 \text{ ft}$$

The resultant design drift loading on the unheated storage space is shown in **Figure VII-11.**

7.6   Example 7.4: Roof Step Drift, Low Ground Snow Load

Use the same parameters as Ex. 7.1, with p_g = 15 psf.

Solution

Balanced Loads: For the upper level roof, the values of C_e = 0.9 and $C_t = C_s =$ I = 1.0 are still valid. Thus, for the ground load of 15 psf, the balanced snow load on the upper level roof is

$$p_s = 0.7 C_e C_t C_s I p_g$$
$$= 0.7(0.9)(1.0)^3(15 \text{ psf})$$
$$= 9.5 \text{ psf; round to } 10 \text{ psf}$$

The minimum roof load from Section 7.3 is $I p_g = 1.0(15 \text{ psf}) = 15 \text{ psf}$.

For the lower level roof, the values of $C_e = 1.0$, $C_t = 1.2$, $C_s = 1.0$, and $I = 1.0$ are still valid. Thus, for the ground snow load of 15 psf, the balanced load on the lower level roof is

$$p_s = 0.7 C_e C_t C_s I p_g$$
$$= 0.7(1.0)(1.2)(1.0)(1.0)(15 \text{ psf})$$
$$= 12.6; \text{ round to } 13 \text{ psf}$$

Drift Loads: The balanced snow depth on the lower level roof is determined from p_s and the snow density (Eq. (7-4)):

$$h_b = \frac{p_s}{\gamma} = \frac{13 \text{ psf}}{0.13(15 \text{ psf}) + 14 \text{ pcf}} = 0.82 \text{ ft}$$

The surcharge height for the leeward drift (wind from the north) is

$$h_d = 0.43 \sqrt[3]{\ell_u} \sqrt[4]{p_g + 10} - 1.5$$
$$= 0.43(100 \text{ ft})^{\frac{1}{3}}(15 \text{ psf} + 10)^{\frac{1}{4}} - 1.5$$
$$= 2.96 \text{ ft}$$

while the corresponding value for the windward drift (wind from the south) is

$$h_d = 0.75 \left[0.43 \sqrt[3]{\ell_u} \sqrt[4]{p_g + 10} - 1.5 \right]$$
$$= 0.75 \left[0.43(170)^{\frac{1}{3}}(15 \text{ psf} + 10)^{\frac{1}{4}} - 1.5 \right]$$
$$= 2.87 \text{ ft}$$

As with Ex. 7.1, the leeward drift height is larger. Since the leeward drift height is less than the clear height ($h_d = 2.96 \text{ ft} < h_c = 10.0 \text{ ft} - 0.82 \text{ ft} = 9.18 \text{ ft}$), the width is equal to four times the surcharge height:

$$w = 4 h_d = 4(2.96 \text{ ft}) = 11.8 \text{ ft}$$

and the drift surcharge load is

$$p_d = h_d \gamma = (2.96 \text{ ft}) \left[(0.13 \times 15 \text{ psf}) + 14 \text{ pcf} \right] = 47.2 \text{ psf}$$

Because the ground snow load is comparatively small ($p_g \leq 20$ psf), minimum roof snow loads (Section 7.3) and the rain-on-snow surcharge need to be considered.

Notice that the rain-on-snow augmented design load and the minimum roof snow load correspond to a uniform or balanced load case and hence

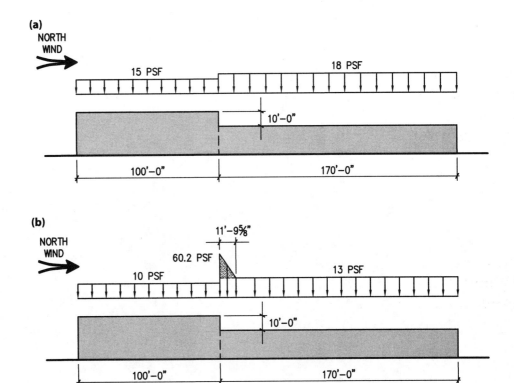

Figure VII-12 Two Load Cases for Step Roof in Ex. 7.4. (a) Uniform Load Case. (b) Balanced Plus Drift Load Case.

do not need to be used in combination with drift, sliding, unbalanced, or partial loads.

The minimum roof snow load for both the upper and lower roofs is Ip_g or 15 psf. The rain-on-snow surcharge of 5 psf is added to the sloped roof loads. For the upper level roof, the sloped roof load is 10 psf and the rain-on-snow surcharge is 5 psf, for a total of 15 psf. For the lower level roof, the sloped roof load is 13 psf plus 5 psf rain-on-snow, for a total of 18 psf. For both roof levels, the sum of the rain-on-snow surcharge and the sloped roof snow load is greater than or equal to the minimum roof snow load; therefore, the rain-on-snow augmented load governs.

It is unclear which load case governs for the lower level roof, so both the uniform load case and the balanced-plus-drift load case require evaluation. The resulting design load cases are shown in **Figure VII-12.**

8 Roof Projections

Snow drifts frequently form at parapet walls and adjacent to rooftop equipment. These drifts are considered windward drifts because the drift forms upwind of the wall or rooftop unit (RTU) and the snow source is located on the upwind roof as opposed to snow originating from the top of the RTU or the parapet wall. As such, these roof projection drifts follow the same provisions as windward roof step drifts discussed in Chapter 7 (Section 7.7 of ASCE 7-02). The drift height is taken as three-quarters of the value given by **Eq. (VII-3),** where ℓ_u is the roof fetch distance upwind of the roof projection.

In a review of snow drift case histories, O'Rourke and DeAngelis (2002) demonstrated that the three-quarters factor applied to windward drifts is reasonable. The observed surcharge drift heights for six windward drifts were compared with values predicted by the ASCE 7-02 provisions. The resulting graph is presented in **Figure VIII-1.**

In one case, the observed height of 2 ft filled the space available for drift formation. If the parapet wall had been taller, then a larger drift may have formed. This full-drift situation is shown by a horizontal line with question marks located to the right-hand side. In another case history, the observed surcharge height was generally characterized as a range from 2.5 ft to 3.5 ft.

For three of the case histories, the provisions over-predict the observed values while the provisions under-predict two observed values. The ratios of observed to predicted range from 0.53 to 1.23 with a mean of 0.84. For the windward drifts considered, the overload for the surcharge was no more than 23%. As noted by O'Rourke and DeAngelis, the overload for the total snow load (balanced plus surcharge) is less, and it is unlikely that a snow

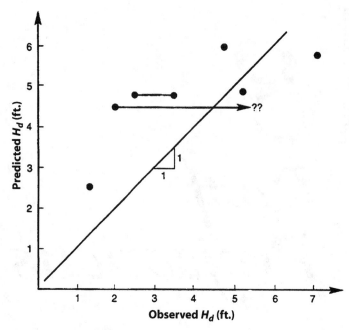

Figure VIII-1   Comparison of Observed Windward Surcharge Drift Height with Values Predicted by ASCE 7-02

overload of 23% would result in significant structural performance problems given the safety factors commonly used in building design.

As with drifts on lower roofs discussed in Chapter 7, the sloped roof snow load in Eq. (7-2) is the balanced snow load below the roof projection drift load. The sloped roof snow, p_s, is $0.7 C_e C_t C_s I p_g$, where p_g is the 50-yr mean recurrence interval (MRI) ground snow load. Minimum loading and rain-on-snow surcharge loads do not influence this balanced load. Finally, if the length of the roof projection (the plan dimension perpendicular to the direction of wind under consideration) is less than 15 ft, then the drift load does not need to be considered along that length. Drifts will form for smaller lengths, but the crosswind plan dimension of the drift and the total drift load (in pounds) is relatively small and can be neglected without impacting the overall integrity of the structural system. The authors are not aware of any structural performance problems related to this 15-ft cutoff for roof projection drifts.

8.1   Example 8.1: Parapet Wall Drift

Determine the design snow drift loads for the roof structure shown in **Figure VIII-2.** The site is in a suburban area (Terrain Category B) where $p_g = 30$ psf. There is a line of conifers about 50 ft to the west of the structure that serves as an obstruction (i.e., top of trees more than 5 ft, 50 ft/10, above roof elevation). The structure is a large heated warehouse deemed to be of ordinary importance. It has parapet walls on the east and west elevations

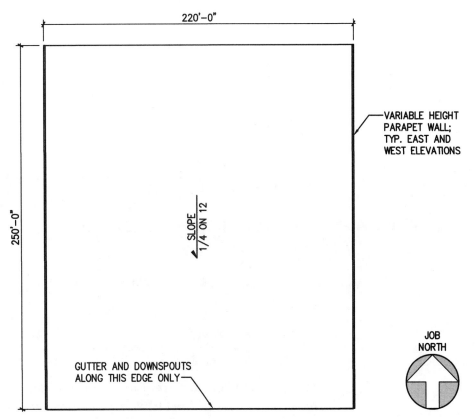

Figure VIII-2   Plan of Monoslope Roof for Ex. 8.1

only. The parapet wall is nominally flush with the roof edge at the north elevation.

<table>
<tr><td>**Solution**</td><td>**Balanced Load:** The building is located in Terrain Category B and the roof is partially exposed (due to the presence of the conifers as well as the parapet wall); therefore, $C_e = 1.0$ from Table 7-2. From Tables 7-3 and 7-4, $C_t = I = 1.0$. For a roof slope of ¼ on 12, $C_s = 1.0$ irrespective of the roof's surface or thermal characteristics. Hence, the balanced load is</td></tr>
</table>

$$p_s = 0.7\,C_e\,C_t\,C_s\,I\,p_g$$
$$= 0.7\,(1.0)\,(1.0)\,(1.0)\,(1.0)\,(30 \text{ psf})$$
$$= 21 \text{ psf}$$

Drift Load: The height of the parapet wall at the southeast and southwest corners is

$$h = 250 \text{ ft}\left(\frac{\frac{1}{4} \text{ in.}}{\text{ft}}\right)$$
$$= 62.5 \text{ in.}$$
$$= 5.2 \text{ ft}$$

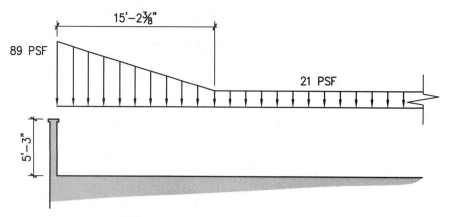

Figure VIII-3 Parapet Wall Drift at Southwest Corner for Ex. 8.1

The snow density, γ, is 0.13(30 psf) + 14 = 18 pcf (Eq. (7-4)), and the depth of the balanced snow is

$$h_b = \frac{p_s}{\gamma} = \frac{21 \text{ psf}}{18 \text{ pcf}} = 1.17 \text{ ft}$$

The space available for drift formation (the clear height above the balanced snow, h_c, is 5.2 − 1.17 = 4 ft) is large compared to the balanced snow depth ($h_c/h_b > 0.2$); therefore, the parapet wall drift must be considered. For an upwind fetch of 220 ft and a ground snow load of 30 psf,

$$h_d = \tfrac{3}{4}\left[0.43(220)^{\frac13}(30+10)^{\frac14} - 1.5\right] = 3.8 \text{ ft}$$

The drift height is not limited by the space available for drift formation because the drift height is less than h_c. The maximum drift surcharge load at the parapet wall is

$$p_d = \gamma h_d = 18 \text{ pcf } (3.8 \text{ ft}) = 68 \text{ psf}$$

and the total maximum load (balanced plus drift) is 21 psf + 68 psf = 89 psf, while the lateral extent is

$$w = 4h_d = 4(3.8 \text{ ft}) = 15.2 \text{ ft}$$

The resulting parapet wall drift load at the southwest corner is shown in **Figure VIII-3.** The design drift at the southeast corner is similar. Since both drifts have the same snow source area, it is unlikely that both design drifts would occur simultaneously. The issue of the possible simultaneous occurrence of drifts adjacent to an RTU is discussed in Chapter 12 (Question 2).

The ground snow load in this case is large enough that the minimum load ($p_f = I \times 20$ psf or 20 psf) is less than the balanced load and does not govern. Similarly, the ground load is large enough that the rain-on-snow surcharge does not apply (see Section 7.10).

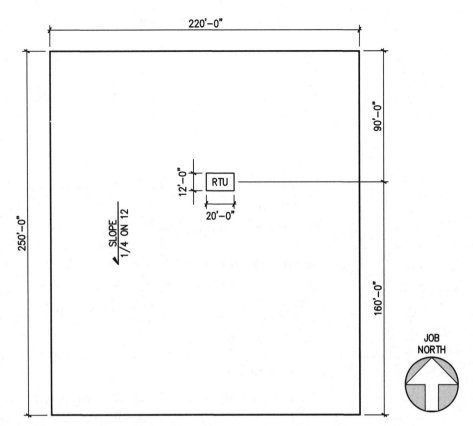

Figure VIII-4 Plan View of Monoslope Roof for Ex. 8.2

If the roof is a continuous-beam system (e.g., a metal building system roof with lapped purlins), then the roof also needs to be checked for the partial loading provisions in Section 7.5. The resultant partial loading would be considered a separate loading case from the balanced-plus-drift loading case determined above.

8.2 Example 8.2: Roof Top Unit (RTU) Drift

Same as Ex. 8.1, except a 4-ft × 12-ft × 20-ft RTU is located as shown in **Figure VIII-4** and the roof has no parapets.

Solution

Balanced Load: Although the parapets have been removed, the stand of conifers is still in close proximity and hence C_e, C_t, I, and p_f are the same as Ex. 8.1 (p_f = 21 psf and h_b = 1.17 ft).

Drift: The clear height to the top of the RTU is h_c = 4.0 ft − 1.17 ft = 2.8 ft and h_c/h_b > 0.2; therefore, a roof projection drift needs to be considered. For wind out of the south, the upwind fetch is 160 ft; hence,

$$h_d = 0.75[0.43(160)^{\frac{1}{3}}(30+10)^{\frac{1}{4}} - 1.5] = 3.3 \text{ ft}$$

Since this drift height is greater than the clear height, h_c, the drift width, w, is larger than $4 h_d$.

Using the "equating the areas" relation from Section 7.7.1, the drift width is

$$w = \frac{4(h_d)^2}{h_c} = \frac{4(3.3 \text{ ft})^2}{2.8 \text{ ft}} = 15.6 \text{ ft}$$

Yet from the "aerodynamically streamlined drift" relation, the drift width cannot exceed

$$w \leq 8 h_c = 8(2.8) = 22.4 \text{ ft}$$

In this case, the "equating the area" relation controls and the total maximum load (balanced plus drift) is

$$p_{max} = h_{RTU} \times \gamma = 4.0 \text{ ft}(18 \text{ pcf}) = 72 \text{ psf}$$

$$p_s = 21 \text{ psf}$$

$$w = 15.6 \text{ ft}$$

For wind out of the north, a smaller windward drift would form adjacent to the RTU because the upwind fetch is only 90 ft. Note that winds from the east or west are not considered because the across-wind side dimension of the RTU is equal to 12 ft, which is less than the 15-ft minimum.

Note: Drifting is frequently a problem with a new or enlarged RTU on an existing roof. Since reinforcing an existing roof is often complicated and expensive, it might be desirable to raise the base of the replacement RTU high enough above the roof level so that windward drifts do not form. An example is described in Chapter 12.

8.3 Example 8.3: Parapet Wall Drift, Low Ground Snow Load

Same as Ex. 8.1, except the ground snow load, p_g, is 15 psf.

Solution   **Balanced Load:** In Ex. 8.1, $p_g = 30$ psf and the balanced snow load is 21 psf for the structure. The balanced snow load is proportional to the ground snow load; therefore, the new balanced load is

$$p_s = \frac{15 \text{ psf}}{30 \text{ psf}}(21 \text{ psf}) = 10.5 \text{ psf, round to 11 psf}$$

Of course, the new balanced load also could have been calculated directly from Eq. (7-2). Recalling from Ex. 8.1 that $C_e = C_t = C_s = I = 1.0$,

$$p_s = 0.7 C_e C_t C_s I p_g$$

or

$$p_s = 0.7(1.0)^4 (15 \text{ psf})$$
$$= 10.5 \text{ psf; round to } 11 \text{ psf}$$

The roof geometry has not changed, and the parapet wall height, h, at the southeast and southwest corners is still 5.2 ft.

The new snow density, γ, is 0.13(15) + 14 = 15.9 pcf (round to 16 pcf), and the depth of the balanced load below the parapet wall drift becomes

$$h_b = \frac{p_s}{\gamma} = \frac{11 \text{ psf}}{16 \text{ pcf}} = 0.69 \text{ ft; round to } 0.7 \text{ ft}$$

Therefore, enough space is available (h_c = 5.2 ft – 0.7 ft = 4.5 ft) for formation of a significant drift ($h_c / h_b > 0.2$).

For the upwind fetch of 220 ft and the ground snow load of 15 psf,

$$h_d = \tfrac{3}{4}\left[0.43(220)^{\frac{1}{3}}(15+10)^{\frac{1}{4}} - 1.5 \right] = 3.2 \text{ ft}$$

Note that the new surcharge drift height is less than that for Ex. 8.1, but not significantly less: it is only 84% of the previous value of 3.8 ft. Although h_d is an increasing function of p_g, the increase is less than a one-to-one ratio.

As in Ex. 8.1, the drift height is less than h_c (3.2 ft < 4.5 ft). Hence, the surcharge height is not limited by the space available for drift formation, and the width or lateral extent from the parapet is four times the surcharge height.

$$w = 4h_d = 4(3.2 \text{ ft}) = 12.8 \text{ ft}$$

The maximum drift surcharge load is

$$p_d = h_d \gamma = 3.2 \text{ ft } (16 \text{ pcf}) = 51.2 \text{ psf; round to } 51 \text{ psf}$$

Thus, the total maximum load (balanced plus drift surcharge) at the parapet wall is 11 psf + 51 psf = 62 psf. The resulting load at the southwest corner is shown in **Figure VIII-5.**

For this example, the ground snow load is small enough that the minimum load or the rain-on-snow enhanced uniform load may govern. Finally, if the roof is a continuous-beam system, various partial loading cases also must be checked.

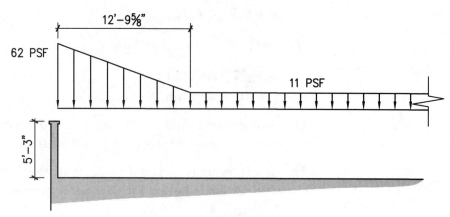

Figure VIII-5 Parapet Wall Drift at Southwest Corner for Ex. 8.3

9

Sliding Snow Loads

As explained in Chapter 4, there are theoretical differences between the design snow load on a nominally flat roof and on a sloped roof. On sloped roofs, snow simply slides off, or for very steep slopes, it does not stick in the first place. From a structural standpoint, snow sliding off a roof is beneficial as long as the sliding snow does not collect in an undesirable location. The roof geometry and the immediate adjacent site plan should be such that the snow sliding off a roof does not pose a hazard to people, parked cars, or other adjacent objects. Clever designers in snowy climates often locate the main entrance at an end wall (e.g., a north or south wall for a north-south ridgeline) of a gable roof structure to avoid snow sliding onto people. If a main entrance is located along a side wall (east or west wall for a north-south ridgeline), then the designer often places a small gable roof above the entrance to deflect sliding snow to either side. This small cross-gable roof, however, can lead to large ice dams.

Snow that slides off a roof and collects against a wall is another concern. In this instance, the snow pile exerts a lateral load on the wall. Some metal building manufacturers offer snow girts as an option for such situations. ASCE 7-02 does not address this issue; however, Chapter 12 of this guide offers suggestions for estimating the load approximately. ASCE 7-02 does, however, have design load provisions for snow that slides onto an adjacent roof. These are discussed below.

At first glance, one might think that the load, which slides onto a lower roof, should be the complement of the sloped roof load, p_s, and the sliding load plus the sloped roof snow load should equal the flat roof load, p_f. If this were the case, the sliding load on the lower roof would be proportional to $(1 - C_s)$, where C_s is the slope factor for the upper roof. This approach appears to be compatible with physics and makes sense intuitively. Following the $(1 - C_s)$ approach, low sloped upper roofs would produce small sliding loads, and steeply sloped upper roofs would produce large sliding loads. The following example explains why this theory is flawed.

Consider a case when the only significant snowfall for a winter season occurs on February 1, resulting in 15 psf of snow on a sloped roof. The weather remains cold and cloudy for the next few days and then it becomes warm and sunny. Upon the arrival of the mild conditions, the upper roof snow begins to melt, and it all slides en-masse onto the lower level roof on February 6. In this case, the sliding snow load is proportional to the sloped roof load. That is, the annual maximum snow load on the upper roof of 15 psf occurred between February 1 and February 6. The sliding load on the lower roof, which arrives on February 6, was due to the same 15 psf originally on the upper roof. The $(1 - C_s)$ principle is flawed because it neglects the aspect of time; an extant upper roof snow might be sliding load snow in the future.

The above example could lead to an "equality" concept, whereby the sliding snow load on the lower roof is proportional to the sloped roof factor, C_s, for the upper roof; however, this reasoning also is flawed. Consider a steep roof subject to a number of snowfalls over the course of a winter. Each snowfall initially sticks to the steeply sloped roof, but its stability is precarious and eventually a gust of wind or a slamming door causes the upper roof snow to slide onto a lower roof. In this case, the load on the upper roof is never very large, but the accumulated sliding load could be substantial. The "equality" concept is flawed because more than one sliding event may occur over the course of a winter season, and the design snow load for the steep upper roof is small compared to the accumulated sliding load on the lower roof.

Because there is not sufficient case history information to establish a more detailed approach that includes the sloped roof factor, C_s, ASCE 7-02 prescribes a simple approach in Section 7.9. The total sliding load per unit length of eave should be $0.4p_fW$, where W is the horizontal distance from the eave to ridge for the sloped upper roof. This sliding snow load is distributed uniformly on the lower level roof over a distance of 15 ft starting from, and perpendicular to, the upper roof eave. If the horizontal measurement of the lower roof from the eave of the upper roof to the edge of the lower roof is less than 15 ft, the sliding load is reduced proportionately.

Recognizing that the potential for sliding snow is an increasing function of roof slope, ASCE 7-02 provides lower bounds where sliding loads do not need to be considered. For instance, as shown in **Figure IX-1,** sliding only occurs when the component of the gravity load parallel to the roof surface (proportional to sin θ) is larger than the frictional resistance (proportional to cos θ). These lower limits for sliding snow are ¼-on-12 for slippery roof surfaces and 2-on-12 for other (nonslippery) surfaces. These lower limits are approximately half the slope for some case histories where sliding

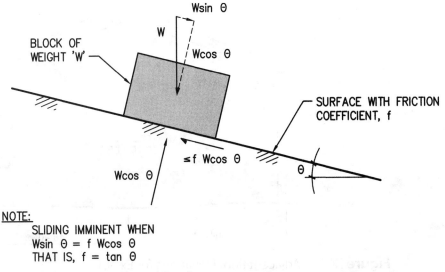

NOTE:
SLIDING IMMINENT WHEN
Wsin θ = f Wcos θ
THAT IS, f = tan θ

Figure IX-1 Onset of Sliding on a Sloped Roof

snow was known to have occurred; sliding has been recorded on a slippery ½-on-12 roof and on a nonslippery 4-on-12 roof.

It is reasonable to assume that these limits and the sliding load are related to the thermal factor, C_t, for the upper roof. With all other things being equal, the potential for snow sliding off a warm roof is greater than for snow sliding off a cold roof. Such refinement of sliding snow loads requires additional case history information.

Finally, the sliding snow load is superimposed on the lower roof's balanced load. The sliding snow load may be reduced if a portion of the snow from the upper roof is blocked by any combination of balanced and/or sliding snow on the lower roof. As with partial loading and the balanced load below a drift, the balanced load on the lower roof for the sliding load case is p_s, as given in Eq. (7-2). Therefore, the sliding load from the upper level roof is superimposed on $0.7 C_e C_t C_s I p_g$ for the lower roof.

9.1 Example 9.1: Sliding Snow Load, Residential Gable Roof (4 on 12)

Determine the design roof snow load due to sliding for an unheated garage attached to a "cold roof" (heated but also vented), shingled residence as sketched in **Figure IX-2**. The structures are located in a suburban site (Terrain Category B) with scattered, nearby tall trees and $p_g = 30$ psf.

| Solution | **Flat Roof and Balanced Loads:** Both the residence and the garage are partially exposed (trees provide some shelter for residence, and trees and residence provide some shelter for garage), and the building is in Terrain Category B; thus, $C_e = 1.0$ from Table 7-2. From Table 7-3, $C_t = 1.1$ for the cold roof residence and $C_t = 1.2$ for the unheated garage. The residence is con- |

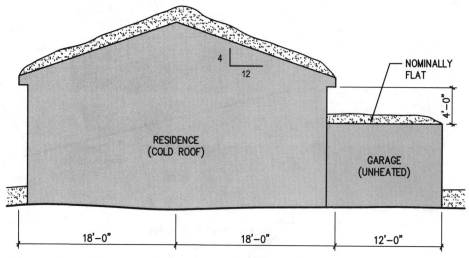

Figure IX-2 Adjacent Roof Elevations for Ex. 9.1

sidered to be of ordinary importance ($I = 1.0$). The garage is considered to be a Category I structure from Table 1-1 because it is a minor storage facility; therefore, from Table 7-4, $I = 0.8$.

For the residence, the flat roof load upon which the sliding surcharge is based is

$$p_f = 0.7 C_e C_t I p_g$$
$$= 0.7(1.0)(1.1)(1.0)(30 \text{ psf})$$
$$= 23 \text{ psf}$$

Since the garage roof is nominally flat, $C_s = 1.0$ irrespective of the roof material and its slipperiness. Hence, the balanced load on the garage roof is

$$p_s = 0.7 C_e C_t C_s I p_g$$
$$= 0.7(1.0)(1.2)(1.0)(0.8)(30 \text{ psf})$$
$$= 20 \text{ psf}$$

Sliding Load: The upper roof is nonslippery (shingles) and has a slope larger than 2-on-12; therefore, sliding snow must be considered. The sliding load is

$$S_L = 0.4 \, p_f W = 0.4(23 \text{ psf})(18 \text{ ft}) = 166 \text{ plf}$$

Since the garage is only 12 ft wide, which is less than the prescribed 15-ft tributary width for sliding snow, the sliding load is reduced by the ratio of 12/15, giving

$$(S_L)_{\text{reduced}} = (166 \text{ plf})12/15 = 133 \text{ plf}$$

This reduced sliding load is uniformly distributed over the 12-ft garage width, resulting in

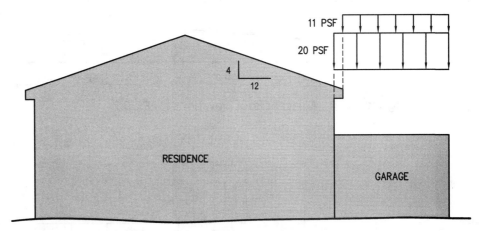

Figure IX-3 Garage Roof Design Snow Load for Ex. 9.1

$$(S_L)_{\text{equiv}} = \frac{133 \text{ plf}}{12 \text{ ft}} = 11.1 \text{ psf}$$

which is equivalent to 166 plf distributed over 15 ft. Hence, the total snow load for the garage (sliding surcharge plus balanced) is 11 psf + 20 psf = 31 psf. The snow density for p_g = 30 psf is γ = 0.13(30 psf) + 14 = 18 pcf, and the total snow depth (balanced plus surcharge) is 31 psf/18 pcf = 1.7 ft. This total depth is less than the 4 ft available; thus, the sliding snow surcharge is not reduced because of blockage. The design roof load for the garage is as shown in **Figure IX-3.**

9.2   Example 9.2: Sliding Snow Load, Commercial Gable Roof (1 on 12)

Determine the design roof snow load due to sliding for the lower heated space shown in **Figure IX-4.** Assume the upper roof surface is standing seam metal, $C_t = C_e = I = 1.0$ for both roofs, and p_g = 35 psf.

Solution

The flat roof snow load for both the upper and lower portions is

$$p_f = 0.7 C_e C_t I p_g$$
$$= 0.7(1.0)^3 (35 \text{ psf})$$
$$= 25 \text{ psf}$$

Since the lower roof is nominally flat, $C_s = 1.0$ and $p_s = C_s p_f = 25$ psf for the lower portion.

Sliding Load: The upper roof is slippery with a slope greater than ¼ on 12; therefore, sliding loads need to be considered. The sliding load is

$$S_L = 0.4 \, p_f W = 0.4(25 \text{ psf})(120 \text{ ft}) = 1{,}200 \text{ plf}$$

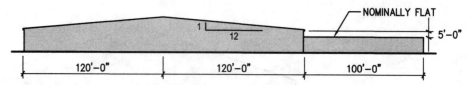

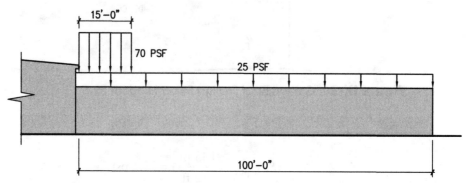

Figure IX-4 Adjacent Roof Elevations for Ex. 9.2

Figure IX-5 Lower Roof Design Snow Load for Ex. 9.2

and when the linear load is distributed over a 15-ft width, the sliding load becomes

$$(S_L)_{equiv} = 1,200 \text{ plf}/15 \text{ ft} = 80 \text{ psf}$$

The total depth corresponding to the balanced load on the lower roof plus the surcharge load from the upper roof (25 psf + 80 psf = 105 psf) requires knowledge of the snow density. For p_g = 35 psf, γ = 0.13(35 psf) + 14 = 19 pcf. Hence, the total depth for 105 psf is

$$h_{tot} = \frac{105 \text{ psf}}{19 \text{ pcf}} = 5.5 \text{ ft}$$

Therefore, the sliding snow is blocked because the calculated snow depth on the lower roof (5.5 ft) exceeds the storage space available (5 ft). A fraction of S_L is forced to remain on the upper roof. Based on the 5-ft-high storage space, the total load on the lower roof near the upper roof eave is

5.0 ft (19 pcf) = 95 psf

and the loading is as shown in **Figure IX-5.**

Note that the windward roof step drifting on the lower roof for the 100-ft upwind fetch for wind from the right also needs to be checked.

9.3   Example 9.3: Sliding Load, Low Ground Snow Area

Same as Ex. 9.1, but for a site with $p_g = 15$ psf.

Solution

For the residence, we now have a flat roof load of

$$p_f = 0.7 C_e C_t I p_g$$
$$= 0.7(1.0)(1.1)(1.0)(15 \text{ psf})$$
$$= 11.6 \text{ psf; round to 12 psf}$$

while for the garage, the sloped roof load is

$$p_s = 0.7 C_e C_t C_s I p_g$$
$$= 0.7(1.0)(1.2)(1.0)(0.8)(15 \text{ psf})$$
$$= 10.1 \text{ psf ; round to 10 psf}$$

Because of the slope of the residence roof, sliding needs to be considered and the surcharge load on the garage is

$$(S_L)_{equiv} = \frac{0.4(12 \text{ psf})(18 \text{ ft})}{15 \text{ ft}} = 5.8 \text{ psf; round to 6 psf}$$

As demonstrated in Ex. 9.1, note that the short lower roof width (12 ft < 15 ft tributary width) is automatically accounted for when the sliding surcharge is characterized as a pressure (with units of psf).

The total load of sliding plus balanced load on the garage is 6 psf + 10 psf or 16 psf total.

Due to the comparatively low ground snow load and the nominally flat slope for the garage roof, additional checks for a uniform load based on minimum roof snow loads (Section 7.3) and rain-on-snow surcharge loads (Section 7.10) are required for the garage roof.

For the site with $p_g = 15$ psf and a garage with $I = 0.8$, the minimum roof snow load per Section 7.3 is

$$p_f = I \times p_g = 0.8(15 \text{ psf}) = 12 \text{ psf}$$

The rain-on-snow surcharge of 5 psf is added to the garage flat roof load of 10 psf, yielding a total uniform load of 15 psf.

The sliding load plus balanced load of 16 psf is greater than both the minimum roof load of 12 psf and the rain-on-snow augmented load of 15 psf, and it extends over the whole garage width of 12 ft; therefore, the sliding surcharge load case governs. The governing garage roof load is shown in **Figure IX-6.**

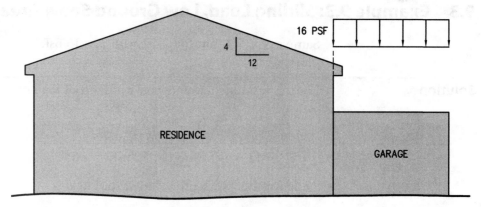

Figure IX-6   Governing Garage Roof Design Snow Load for Ex. 9.3

10

Rain-on-Snow Surcharge Loads

It is not unusual for it to rain while a building's roof is still covered with snow. This can occur, for example, in spring before the winter snowpack has melted completely. The rain-on-snow provisions in Section 7.10 of ASCE 7-02 are intended to cover such loading situations.

A rain-on-snow surcharge is occasionally one of the contributing factors in a roof collapse although there are no known case histories wherein rain-on-snow, of and by itself, was the sole source of an overload that led to a collapse. For example, a series of mixed precipitation events over the 1996–1997 holiday season led to a large number of structural collapses in the Pacific Northwest. According to a 1998 report by the Structural Engineers Association of Washington (SEAW), a total of 1,663 buildings were damaged by the holiday storms.

Analysis of three of these 1996–1997 Pacific Northwest collapses—one located in the mountains of central Washington and another two located in the greater Yakima area—showed that, in all three cases the structures were cold rooms (intentionally kept at or below freezing), with measured roof snow loads greater than the design roof snow load (57 psf versus 32 psf, 29 psf versus 22 psf, and 36 psf versus 30 psf). In all three cases, other design issues were discovered with the roof that were unrelated to the snow load. In all of these cases, calculations suggest that the contribution of rain on snow to the total load was modest. In two of the buildings, the surcharge at the eave for a rain-on-snow scenario or for the rain simply refreezing in the roof snowpack was about 5% of the total load, and in the third building, the roof collapsed prior to the rain.

Four possible rain-on-snow scenarios are outlined in **Table X-1** and described below.

Table X-1 Rain-on-Snow Scenarios

Scenario	Roof Snowpack	Ground Snowpack	Comment
1	Rain refreezes	Rain refreezes	p_g presumably includes rain
2	Rain flows through	Rain refreezes	p_g presumably includes rain
3	Rain refreezes	Rain flows through	Covered by $C_t = 1.2$??
4	Rain flows through	Rain flows through	Covered by Section 7.10

- **Scenario 1:** The rain refreezes within both the ground and roof snowpacks. This is most likely to occur if both snowpacks are cold and comparatively deep. In this case there is a semipermanent increase in both the roof and ground snow loads. Presumably this additional load would be recorded as part of any ground snow load measurement program.
- **Scenario 2:** The rain refreezes in the ground snowpack but percolates through the roof snowpack. This is most likely to occur if the roof snowpack is warmer or shallower than the ground snowpack, as would be expected if the building were heated. In this case, there is a semipermanent increase in the ground load (as in Scenario 1) but a transitory increase in roof load as the rain percolates down through the roof snowpack to the roof surface and flows down to the eaves or drains.
- **Scenario 3:** The rain percolates through the ground snowpack but refreezes within the roof snowpack. This is most likely to occur if the roof snowpack is colder and/or deeper than the ground snowpack. This could occur if the space immediately below the roof layer is intentionally kept at or below freezing (e.g., a cold room). In this case, there is a semipermanent increase in the roof load but only a transitory increase in the ground load. It seems unlikely that this temporary increase in the weight of the ground snowpack would be recorded as part of a ground load measurement program.
- **Scenario 4:** The rain percolates through both the roof and ground snowpacks. This is more likely to occur if both of the snowpacks are warm and comparatively shallow. In this case, there is a transitory increase in both the roof and the ground snowpack weight. However, as with Scenario 3, it is unlikely that a ground load measurement program would capture this temporary surcharge.

In Scenarios 1 and 2, the rain refreezes in the ground snowpack and will likely be captured in ground snow load measurements. Hence, there is no need to add a special surcharge to account for rain-on-snow effects

Table X-2 Estimated Roof Rain-on-Snow Surcharge for ¼-on-12 Roof Slope and 100-ft Eave-to-Ridge Distance

Method	Range (psf)	Average (psf)
50-yr (wintertime) daily rain in 19 cities	0 to 5.68	2.02
2-yr (year-round) rain in 19 cities	1.03 to 7.61	4.73
1996–1997 winter storm: Seattle and Yakima	—	4.60

Source: Adapted from O'Rourke and Downey (2001).

because the effects are accounted for in the 50-yr ground snow load and thus in the design roof snow loads as well. This assumes that the semipermanent load due to the rain refreezing in the roof snowpack (Scenario 1) is comparable to the roof load due to rain flowing downslope in a saturated layer at the base of the roof snowpack (Scenario 2).

In Scenarios 3 and 4, a ground snow load measurement program would not capture the rain flow through the ground snowpack. Because the 50-yr ground snow load does not account for the effects of the rain in these scenarios, the code provisions must attempt to quantify the transitory increase in roof load.

Scenario 4 (i.e., rain percolating through both the roof and ground snowpacks) is covered by the rain-on-snow provisions in Section 7.10 of ASCE 7-02. Scenario 3 (i.e., for cold rooms and freezers where rain refreezes within the roof snowpack but percolates through the ground snowpack) is not addressed specifically by Section 7.10. It is unclear whether the C_t factor of 1.2 for such structures provides adequate coverage for Scenario 3.

The rain-on-snow surcharge associated with rain percolating through a roof snowpack is composed of two portions. The first portion (typically small) is the load caused by vertical percolation through the upper unsaturated layer. The second, larger portion is the load caused by rain flowing downslope in the saturated layer immediately above the roof.

As expected, the surcharge caused by flow in the saturated layer increases with increases in rainfall intensity, rainstorm duration, and eave-to-ridge distance, and decreases as roof slope increases. That is, all other things being equal, the low-sloped roof with a large eave-to-ridge distance will have the greatest surcharge.

Colbeck (1977) developed surcharge load relations for a given rainfall intensity and duration, but it is not possible to determine an exact 50-yr roof surcharge for rain-on-snow due to the lack of appropriate weather information. For example, O'Rourke and Downey (2001) present three estimates of the surcharge for Scenarios 2 and 4 (rain percolating through the roof snowpack). Their results for a ¼-on-12 roof slope (1.19 degrees) and a 100-ft eave-to-ridge distance are presented in **Table X-2.**

Table X-3 Average Roof Surcharge Loads for 1996-1997 Holiday Storm Events in Yakima and Seattle

Eave-to-Ridge Distance W (ft)	Roof Slope (degrees)					
	0.60 (1/8 on 12)	1.19 (1/4 on 12)	2.39 (1/2 on 12)	4.76 (1 on 12)	9.46 (2 on 12)	18.4 (4 on 12)
20	2.53	1.54	1.00	0.81	0.51	0.31
50	4.40	3.38	2.05	1.15	0.89	0.62
100	5.05	4.60	3.56	2.12	1.15	0.90
250	5.28	5.20	4.90	4.06	2.60	1.37

Source: O'Rourke and Downey (2001).

In the first estimate, National Oceanic and Atmospheric Administration (NOAA) ground snow and precipitation data were used to determine the wintertime maximum daily rain for a 50-yr return period. The resulting roof surcharge is only an estimate since an associated rainstorm duration and intensity had to be assumed. The results in **Table X-2** are based on an assumed 1-h duration (i.e., the maximum daily rain was assumed to have fallen over a 1-h period). For the 19 cities considered by O'Rourke and Downey, the roof surcharge loads ranged from 0 psf to 5.68 psf with an average of 2.02 psf.

In the second method, rainfall intensity data were used to determine the 2-yr mean recurrence interval (MRI) rainstorm with a duration of 1 h. The resulting roof surcharge is an upper bound estimate since the 2-yr rain may have occurred when the roof was free of snow. For the 19 cities considered, the roof surcharge loads ranged from 1.03 psf to 7.61 psf with an average of 4.73 psf.

In the third method, the roof surcharge was determined for the 1996–1997 holiday storm rain-on-snow events in Yakima and Seattle. Fairly detailed weather information for Yakima Airport allowed O'Rourke and Downey to quantify the magnitude of the rain-on-snow event. On December 31, 1996, Yakima had a ground snow load of about 23 psf and a rain event with an average intensity of 0.113 in. of rain per hour for a 12-h period. On January 2, 1997, Seattle had a 20-psf ground snow load and a rain event with an average of 0.117 in. of rain per hour for 11 h. Unfortunately, the return period for this holiday storm event is unknown. For the two cities considered, the average roof surcharge load was 4.60 psf.

The influence of roof slope and eave-to-ridge distance is shown for the Yakima and Seattle case histories in **Table X-3**. As one might expect, the roof surcharge is an increasing function of eave-to-ridge distance, W, with the surcharge for W = 250 ft being 3 to 5 times the value for W = 20 ft. Similarly, the surcharge is a decreasing function of roof slope with the value for a ¼-on-12 (1.19-degree) roof slope being 4 to 5 times that for a 4-on-12 (18.4-degree) roof slope.

The rain-on-snow provisions in Section 7.10 require that a 5-psf surcharge be added to the flat roof snow load for roofs with slopes *less* than ½ on 12 (2.39 degrees) in regions where the ground snow load is 20 psf or less. As noted in the Commentary, it is assumed that Scenarios 1 or 2 (rain refreezes in ground snowpack) apply for sites with $p_g > 20$ psf. On the other hand, for lower ground load sites, Scenario 4 (rain percolates through both ground and roof snowpacks) is assumed.

The provisions imply that lower slopes tend to have larger rain surcharge loads. Note, however, that in the ASCE 7-02 Standard the rain-on-snow surcharge is unrelated to the roof size (specifically eave-to-ridge distance). Unfortunately, this leads to inconsistencies. For example, from **Table X-3** the expected surcharge for a 1-on-12 slope with $W = 250$ ft (no design rain-on-snow surcharge required per ASCE 7-02) is actually larger than that for a ¼-on-12 slope with $W = 50$ ft (design surcharge required per ASCE 7-02).

10.1 Example 10.1: Uniform Design Snow Load, Monoslope Roof (¼ on 12)

Same as Ex. 4.1, except for this problem, the slope is ¼ on 12 and $p_g = 15$ psf.

Solution

From Ex. 4.1, $C_e = 1.0$, $C_t = 1.1$, and $I = 1.2$. Also from Figure 7-2, $C_s = 1.0$ for a ¼-on-12 (1.19-degree) slope, irrespective of C_t or any other factor. Hence, the sloped roof snow load is as follows:

$$p_s = 0.7 C_e C_t C_s I p_g$$
$$= 0.7(1.0)(1.1)(1.0)(1.2)(15 \text{ psf})$$
$$= 13.86 \text{ psf; round to } 14 \text{ psf}$$

The slope is less than ½ on 12 and $p_s \leq 20$ psf; therefore, a rain-on-snow surcharge of 5 psf (reference Section 7.10) is added to the sloped roof load of 14 psf to give a uniform value of 19 psf.

The slope of this monoslope roof is less than 15 degrees; therefore, the minimum roof loads must be checked per Section 7.3. Since the ground snow load is less than 20 psf, the minimum is $I p_g = 1.2(15) = 18$ psf.

In this case, the rain-on-snow augmented flat roof load governs, and the design uniform load is 19 psf. Note that for the low slopes where the rain-on-snow surcharge applies, the slope factor, C_s, always equals 1.0; therefore, the flat roof load equals the sloped roof load.

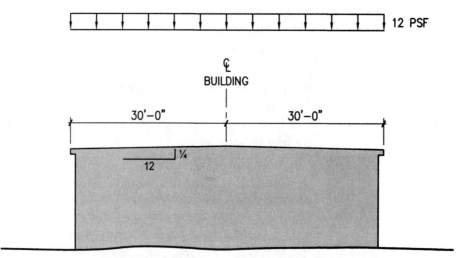

Figure X-1   Uniform Design Load and Elevation of a Gable Roof for Ex. 10.2

10.2   Example 10.2: Uniform Design Snow Load, Gable Roof (¼ on 12)

Determine the uniform design load for the heated symmetric gable roof structure shown in **Figure X-1,** with $C_e = 1.0$, $I = 1.0$, and $p_g = 10$ psf.

Solution

For a heated structure, $C_t = 1.0$ and $C_s = 1.0$ for any roof with a ¼-on-12 slope. Hence, the sloped roof load is

$$p_s = 0.7 C_e C_t C_s I p_g$$
$$= 0.7(1.0)^4 (10 \text{ psf})$$
$$= 7 \text{ psf}$$

Since the roof slope is less than ½ on 12 and the ground snow load ≤ 20 psf, a 5-psf surcharge is applied per Section 7.10. This results in a rain-on-snow augmented load of 12 psf.

Since the ¼-on-12 slope (1.19 degrees) is less than $70/W + 0.5$ [(70/30) + 0.5 = 2.83°], minimum roof loading per Section 7.3.4 must be considered. Since $p_g < 20$, the minimum is Ip_g or 1.0(10) = 10 psf.

In this case, the rain-on-snow augmented load governs and the uniform design load is 12 psf, as shown in **Figure X-1.**

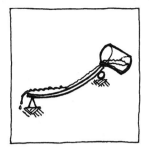

11
Ponding Instability and Existing Roofs

11.1   Ponding Instability

Section 7.11 of ASCE 7-02 requires a ponding analysis for roofs with slopes less than ¼ on 12 and states that the design must use the full snow load in the ponding instability analysis. Section 8.4 of ASCE 7-02 further requires the designer to assume that the primary drainage is blocked in such an analysis.

The Commentary to Sections 7.11 and 7.12 alerts designers of typical problems often encountered with flat or very low-sloped roofs (i.e., less than ¼-on-12), including "unintentional" low spots, increased potential for significant rain-on-snow surcharge, and poor performance of waterproofing membranes. Given the problems associated with flat and very low-sloped roofs, one could argue that the Commentary should simply say, "*Do not design or build with less than a ¼-on-12 roof slope to drainage.*" However, there are some situations, particularly with large roofs, where either a $\frac{1}{8}$-on-12 roof slope or a completely flat roof is desirable to the owner or architect. In such cases, the structural engineer needs to check the roof for ponding, as required by ASCE 7-02.

Section 7.11 implies that a ¼-on-12 roof slope is sufficient to preclude ponding instability. Note that for a beam with a ¼-on-12 slope, any deflection-to-span ratio of 1/100 or less means that the midspan is still above the elevation of the "downslope" support point. As shown in **Figure XI-1,** the

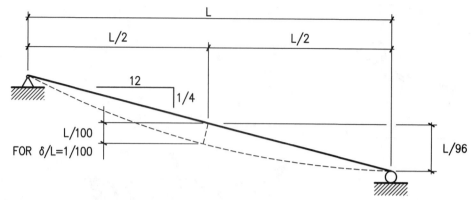

Figure XI-1   Positive Slope to Drainage for ¼-on-12 Roof Slope and δ/L = 1/100 Deflection Criterion

initial elevation difference between the downslope support point and midspan of the straight line beam with a slope of ¼-on-12 is

$$\text{Elevation differences} = \frac{L}{2} \times \frac{\frac{1}{4}}{12} = \frac{L}{96}$$

while the maximum sag at midspan for a δ/L of 1/100 is

$$\text{Max sag} = \frac{L}{100} < \frac{L}{96}$$

In two investigations of ponding failures performed by the authors, the roofs had very low slopes: ⅛ on 12 in one case and apparently dead flat in the other. Also, both had cantilevered girder roof systems with drop-in spans, somewhat similar to that shown in **Figure V-3.**

11.2   Existing Roofs

Section 7.12 requires the designer of a new facility that will be adjacent to an existing lower roof facility to either (a) analyze and possibly reinforce the existing roof for the now-expected drift or sliding loads (if both facilities are part of the designer's scope of work), or (b) make the owner of the existing facility aware of the "new" loads (if the existing facility owner is not the client or if the existing facility is not in the scope of work). ASCE 7-02 and the Commentary do not mention the manner by which the notification is to be made; however, we highly recommend doing this by certified mail, possibly with a copy to the local building official.

The roof step problem for new higher/existing lower roof situations is particularly difficult because of the cost and the complexities of reinforcing an existing roof. Four possible solutions to this problem are discussed in Chapter 12 (Question 6).

12
Frequently Asked Questions

This chapter presents answers to snow loading questions that are not specifically addressed by the ASCE 7-02 provisions. As such, it is *not* an official or unofficial interpretation or extrapolation of the ASCE 7-02 provisions, but rather the authors' opinions on issues often raised in discussions about snow loading.

1. **We have had more snow than usual this winter. We received a call from a new client about an apparent overload of an existing structure. During a site inspection, we noted significant sagging in the roof purlins. What are the options to address this situation?**

The overriding concern here is safety, which means that snow removal from a residential structure with a moderate to steep slope should be accomplished *from the ground* with a "snow rake" (typically available at a local hardware store). In larger nonresidential construction, there are different options. If ponding of water is a significant load for the structure in question, punch a small hole at a low point in the roof to effectively reduce the load. If the roof loading is primarily snow, either use temporary supports or remove snow manually or mechanically. If temporary supports are used, a roof collapse is a safety issue while the temporary supports are being installed. It would be prudent to assign an individual to regularly monitor the roof (looking for an increase in the roof sag) and alert workers if necessary.

For the snow removal option, potential safety issues include workers slipping or sliding off the roof and potential collapse. For structures with continuous-beam systems, the work crew should be dispersed over the roof so that snow is being removed from all spans more or less at the same time. It is undesirable to have the entire crew working on the same span. The snow removal effort should not result in a simulated partial load pattern with full load on some spans and reduced loads on others. Although this may seem obvious, the shoveled or blown snow should land on the ground and not on another portion of the roof being cleared or on a lower roof level.

Finally, we are aware of at least one case where roof snow removal using a fire hose was attempted. This "cleaning the driveway with a garden hose" approach did not prove successful; the structure collapsed during the roof snow removal work.

2. **A client wishes to install a new (8-ft 0-in. × 16-ft 0-in.) rooftop unit (RTU) adjacent to an existing unit. This proposal results in a combined RTU footprint of 16-ft 0-in. square. The roof was initially designed for two simultaneously applied windward drifts, one along each of the 16-ft 0-in.–wide sides of the existing RTU. The roof was not designed for drifts along the 8-ft 0-in. ends because of the 15-ft cutoff given in Section 7.8. The weight of the additional RTU is expected to exhaust most of the excess structural capacity in the existing roof. Is it necessary to apply roof projection drifts to all four of the 16-ft 0-in. sides simultaneously?**

It is clear from Section 7.8 that roof projection drifts need to be considered on all sides of the projection that are at least 15 ft long; hence, for the combined RTU in question, drifts must be considered on all four sides. However, Section 7.8 does not indicate whether the drifts must be considered simultaneously. The question then becomes, is it likely that significant drifting will occur from all four directions at the same site during the same year?

It should be noted that it is common for the wind direction to change even during a single storm event. Low pressure systems move translationally while also rotating about the center of the low (a counter-clockwise rotation is experienced in North America). For example, as a strong low pressure system proceeds from New York City to Boston (depicted in **Figure XII-1**), the wind direction in Albany, N.Y. (located north of New York City and west of Boston) shifts 90 degrees, from easterly to northerly. This pattern is typical for a classic "Nor'easter." At the extreme, a low pressure system that travels directly over a site results in a wind shift of 180 degrees (e.g., shifting from southerly to northerly for a low traveling west to east). These examples suggest that, for a single storm traveling along a nominally straight track, drifts could form along two adjacent sides (90-degree wind shift) or two opposite sides (180-degree wind shift) of a roof projection, but not along all four sides (270-degree wind shift) simultaneously. Furthermore, when drifting occurs on two adjacent sides, the initial drift formation (e.g., along the east side) would likely be reduced by the subsequent wind from the north or south.

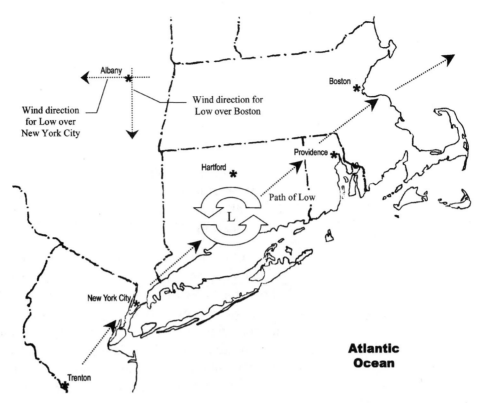

Figure XII-1   Classic Nor'easter—Low Pressure System Traveling from New York City to Boston with a Resulting Wind Shift from Easterly to Northerly at Albany.

Another possibility would be a quartering wind, i.e., a wind out of the southeast for an RTU with adjacent sides facing east and south. Although the wind would hit both faces simultaneously, large drift formations are not likely on either side because of the 45-degree angle of attack in a horizontal plane. One would not expect a large region of aerodynamic shade on either of the two windward faces. If anything, a spike-type drift downwind, trailing off of the northeast and southwest corners, is expected.

Note that it is possible to develop drifts simultaneously on all sides of a roof projection, but it involves a complex scenario that is not probable. If a north to south tracking low travels directly over a site and is followed by a west-to-east low a few days later that again tracks over the site, then the wind would drift the snow from all four compass directions. But in order to prevent erosion of the first set of drifts by the wind from the second storm, an intervening event such as freezing rain or sleet is required that would "freeze" the first set of drifts in place. Although such a scenario is plausible, the return period of this scenario would very likely exceed the 50-yr value envisioned by the ASCE 7-02 snow provisions.

Based on the above, it is reasonable to consider the following loading cases for an existing roof: two separate sets of windward drifts on opposite faces (i.e., north and south as one case, east and west as another) and four separate sets of windward drifts on adjacent faces (i.e., north and east as one case, east and south as another).

For a new design, the marginal cost for a design based on simultaneous drifts on all four sides would likely be minor, and designing for just one load case (all four sides simultaneously), as opposed to the six suggested above, would simplify the design efforts. The "four sides simultaneously" approach would be appropriate if the roof were a new design.

3. **A client wants to install a new RTU on an existing large roof. The existing roof structure can marginally support the proposed RTU but has no additional capacity for drift loads. Is there a means to install the RTU to avoid formation of large windward drifts?**

Commentary Section C7.8 states the following in relation to arrays of rooftop solar collectors: "By elevating collectors several feet (a meter or more) above the roof ... the potential for drifting will be diminished significantly." It is the authors' opinions that conventional RTUs should be elevated such that there is a 2-ft gap between the bottom of the RTU dunnage/ framework and the top of the balanced snow surface. Thus, for the RTU in Ex. 8.2, the RTU should be elevated a minimum of 3.20 ft (balanced + minimum gap; 1.20 ft + 2.0 ft = 3.20 ft) above the roof surface.

If the projection is much taller than an RTU, such as a billboard or cooling tower, then the gap probably should be greater than 2 ft.

4. **What drift load should be used to design the mansard roof sketched in Figure XII-2? The ground snow load is 35 psf and $C_e = C_t = I = 1.0$. The roof is slippery, unobstructed, and unventilated, and it has an R-value of 40 °F h ft^2/BTU.**

The ASCE 7-02 provisions do not address the mansard roof geometry specifically; however, the suggested approach is arguably a logical extension of the provisions that do exist.

The balanced load for the sloping portions of the mansard roof is

$$p_s = 0.7 C_e C_t C_s I p_g$$
$$= 0.7(1.0)^3(.56)(35 \text{ psf})$$
$$= 13.7 \text{ psf}$$

where $C_s = 0.56$ from Figure 7-2a. The upward fetch for the leeward drift on the sloped portion is conservatively taken as 85 ft. It is assumed that the cross-sectional area of the mansard roof drift matches that for a leeward roof step drift with the same fetch and ground load. Hence, for a ground load of 35 psf, the surcharge height of the hypothetical roof step drift is

$$h_d = 0.43\sqrt[3]{85}\sqrt[4]{35+10} - 1.5$$
$$= 3.4 \text{ ft}$$

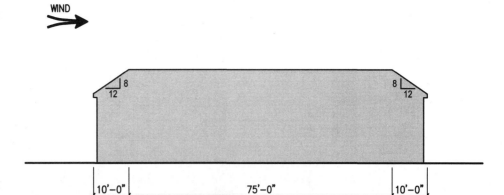

Figure XII-2 Elevation of Mansard Roof for Question 4

The length of this hypothetical step drift is

$$w = 4h_d = 13.6 \text{ ft}$$

This results in a cross-sectional area of the hypothetical step drift (and by assumption the mansard drift) of

$$A = \frac{1}{2}(3.4 \text{ ft})(13.6 \text{ ft}) = 23.1 \text{ ft}^2$$

Furthermore, assuming that the top surface of the mansard roof drift is nominally flat, the triangular drift shape has a vertical depth that is 8/12 (or two-thirds) of its horizontal extent, w_m. Equating the cross-sectional areas and solving for the horizontal extent,

$$23.1 \text{ ft}^2 = \frac{1}{2}(0.66\, w_m)\, w_m$$

Thus,

$$w_m = 8.37 \text{ ft}$$

Therefore, the height of the drift is two-thirds w_m or 5.58 ft. The snow density for $p_g = 35$ psf is $0.13(35) + 14 = 18.6$ pcf, and the surcharge drift load is 5.58 ft (18.6 pcf) = 104 psf. The resulting combined snow load for this mansard roof is shown in **Figure XII-3.**

5. **What is the unbalanced load for the asymmetric gable roof structure illustrated in Figure XII-4? The ground snow load is 30 psf, while $C_e = C_t = I = 1.0$. The roof is unventilated with an R-value of 25 °F h ft^2/BTU.**

The ASCE 7-02 provisions do not give guidance regarding unbalanced loading on asymmetric gable roofs. The Commentary, however, states that the gable roof drift parameter, β, is proportional to the percentage of snow that could drift across the ridgeline, and it notes that, for asymmetric

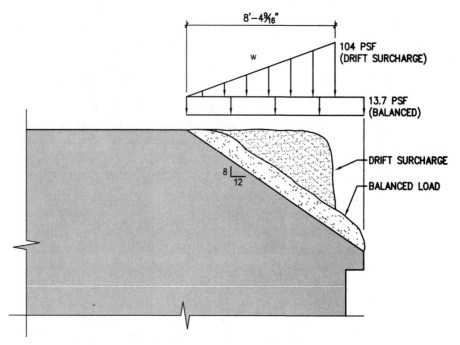

Figure XII-3   Suggested Design Drift Load for Mansard Roof in Question 4

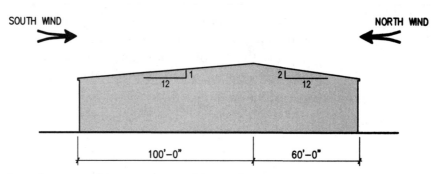

Figure XII-4   East Elevation of Asymmetric Gable Roof in Question 5

gables, the unbalanced load on the side with the smaller W is expected to be larger.

The approach suggested herein assumes that the surcharge is proportional to the size of the snow source area. That is, the β for asymmetric roofs is taken as the value given in Eq. (7-3) (i.e., function of p_g) multiplied by the ratio of windward to leeward widths.

The balanced load (i.e., the sloped roof load) for both sides of the roof is

$$p_s = 0.7 C_e C_t C_s I p_g$$
$$= 0.7(1.0)^4 (30 \text{ psf})$$
$$= 21 \text{ psf}$$

The C_s factor is 1.0 for both portions from Figure 7-2(a) because the thermal environment (unventilated with R < 30) is such that ice dams may form at the eaves.

Both the northerly and the southerly winds must be considered because the roof slopes are larger than the $70/W + 0.5$ limit. For instance, for wind out of the north, the roof slope is 4.76 degrees (1 on 12) while the limit below which unbalanced loading does not need to be considered is $70/100$ ft $+ 0.5 = 1.2$ degrees, or approximately ¼ on 12.

For $p_g = 30$ psf, $\beta = 1.5 - 0.025$ (30 psf) $= 0.75$ from Eq. (7-3). For wind out of the north, the ratio of windward to leeward widths is 60 ft/100 ft and the modified gable roof drift parameter becomes

$$\beta_N = \frac{W_N}{W_S}\beta = 0.6(.75) = 0.45$$

The leeward unbalanced load becomes

$$1.2\left(1+\frac{\beta}{2}\right)\left(\frac{p_s}{C_e}\right) = 1.2\left(1+\frac{.45}{2}\right)\left(\frac{21\ \text{psf}}{1.0}\right)$$
$$= 31\ \text{psf}$$

while the windward load is

$$0.3p_s = 0.3(21\ \text{psf}) = 6.3\ \text{psf}$$

For wind out of the south, the gable roof drift parameter becomes

$$\beta_s = \frac{W_S}{W_N}\beta = \frac{100\ \text{ft}}{60\ \text{ft}}(0.75) = 1.25$$

The leeward load is then

$$1.2\left(1+\frac{\beta}{2}\right)\left(\frac{p_s}{C_e}\right) = 1.2\left(1+\frac{1.25}{2}\right)\left(\frac{21\ \text{psf}}{1.0}\right)$$
$$= 41\ \text{psf}$$

These suggested unbalanced loads for northerly and southerly wind are shown in **Figure XII-5.**

6. **We are planning a new warehouse addition adjacent to an existing facility. The owner requests that the addition have a higher roof elevation than the existing roof. The project budget is tight; therefore, spending money to reinforce the existing roof is not permitted. What are the viable design options?**

This is a very common situation, and four approaches are conceivable.

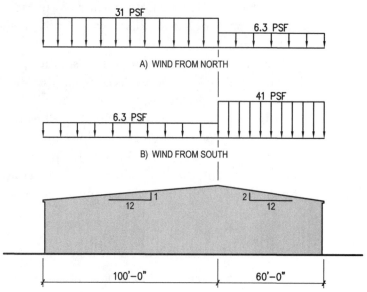

Figure XII-5 Suggested Unbalanced Snow Loads for Asymmetric Gable Roof in Question 5. (a) Wind from North. (b) Wind from South.

Approach 1: Snow Bay

The easiest approach is to design the step, which contains the resulting drift, within a "snow bay" as part of the new addition, as shown in **Figure XII-6.** In this way, the roof step drift load is handled simply and economically by use of stronger members for the portions of the addition now subject to drift loading. The drawback with this approach, of course, is that the snow bay roof elevation must either be the same elevation as the lower existing structure or the roof elevation can only be 1.2 h_b higher than the existing structure (recall from Section 7.7.1 that drift loads are not required if $h_c / h_b < 0.2$).

Approach 2: Reduce Drift Accumulation Space

An alternate approach is to reduce the space available for drift accumulation. This could be done by securing large lightweight foam blocks to the lower level roof or by constructing a false inclined roof between the addition and the existing structure, as shown in **Figure XII-7.** In this manner, much of the aerodynamic shade region at the roof step is spared snow accumulation.

One of the drawbacks with the foam or inclined roof approaches is the extra dead load and partial snow drift loads that the existing facility roof needs to support. The existing columns are the most direct load path, so the framing system for the connector roof could be horizontal purlins spanning parallel to the roof step in plan and supported by inclined beams spanning perpendicular to the step. The inclined beam could then frame into

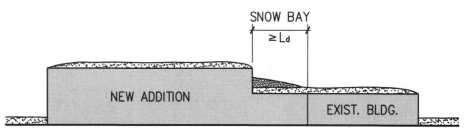

Figure XII-6   New Addition with a Snow Bay

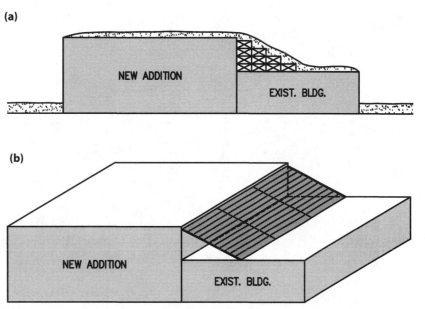

(a)

(b)

Figure XII-7   Reduction of Space Available for Drift Formation. (a) Geofoam Blocks on Existing Roof. (b) Inclined Roof over Drift Accumulation Area.

and be supported by the columns in the existing structure, which may require reinforcement.

Another problem with this approach is that the potential for some sort of drift formation is somewhat unclear. Consider, for example, a fairly typical case wherein the roof elevation difference is 5 ft and the column spacing in the existing facility is 25 ft on-center. This would result in a connector roof slope of 1 on 5. Neglecting the existing roof, the connector roof and the adjacent nominally flat roof for the addition form a mansard-type roof, and unfortunately the ASCE 7-02 provisions do not address drift loads on mansard-type roofs. A rational, but uncodified, method for estimating the potential size of mansard roof drifts is presented in Question 4.

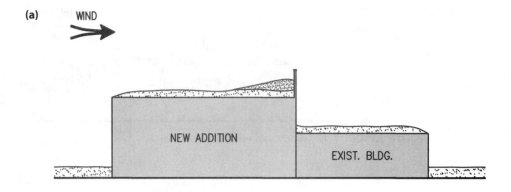

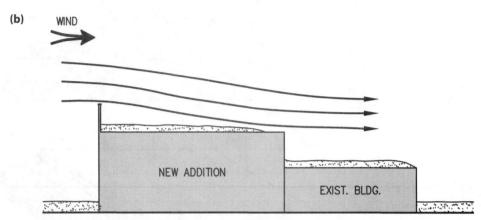

Figure XII-8 Tall Parapet Wall Approaches. (a) Wall at Adjacent/Shared Column Line. (b) Wall at Far End of Addition.

Approach 3: Tall Parapet Wall

A third approach is to introduce a parapet wall that would reduce the size of the leeward roof step drift by either capturing some of the snow on the upper level roof (parapet wall located at the adjacent/shared column line) or by shielding the upper level roof snow from wind (parapet wall located at exterior column line of the new addition), as shown in **Figure XII-8.**

There are two main drawbacks with this approach. First, a parapet wall does not eliminate the potential for a windward drift. In fact, the parapet at the common column line potentially makes things worse, particularly if the existing roof has a large fetch, by providing a larger space for a windward drift accumulation.

Second, a parapet wall of moderate height is not very effective even for a leeward drift formation on the lower roof. Consider a site with a balanced roof snow depth of about 1 ft ($C_e = C_t = I = 1.0$, $p_g = 25$ psf, and $\gamma = 17.3$ pcf), a parapet wall height of 5 ft, and an upper level roof upwind fetch of 150 ft. The clear height of the parapet wall above the balanced snow is then $h_c = h - h_b = 5.0$ ft $- 1.0$ ft $= 4$ ft. The cross-sectional area of a triangular drift with a rise to run of 1 on 4 and a height of 4 ft is

$$\text{Area} = \tfrac{1}{2}h_d w = (0.5)4.0 \text{ ft}\left[4(4.0 \text{ ft})\right] = 32 \text{ ft}^2$$

However, a conservative estimate of the cross-sectional area of the upwind snow source is the depth of the balanced snow times the upwind fetch, or

$$\text{Area} = 1 \text{ ft} \times 150 \text{ ft} = 150 \text{ ft}^2$$

Thus, this moderate-height parapet wall could capture no more than about 21% ($32 \text{ ft}^2/150 \text{ ft}^2 = 0.213$) of the upper roof snow. Hence, the potential for a leeward drift accumulation on the lower roof, due to the "remaining" 79% of snow on the upper roof, still exists. If it were argued that the "full" parapet wall drift would have a rise to run of 1 on 8, the capture area would double to 64 ft^2, or only 42% of the upper roof snow source area, leaving unaccounted for 58% of the upper roof snow.

Even if a parapet wall was built tall enough to provide 150 ft^2 of potential capture area (for a 1-on-4 slope, 8.66 ft + 1.0 ft for the balanced snow results in a 9.66-ft-tall wall), there is no guarantee that a leeward drift would be avoided on the lower level roof. This is because the trapping efficiency of almost any geometric irregularity is less than 100%. That is, all the upper roof snow would not necessarily accumulate at the parapet, even if there were enough space. For example, when the windward drift at the tall parapet is partially full, it forms a snow ramp that would allow some of the saltating snow particles from the upper level roof to jump over the parapet wall, potentially forming a leeward drift on the lower roof.

The difficulty with a parapet wall at the far end of the upper level roof (see **Figure XII-8(b)**) is practicality. For reasonably sized upper roofs, the parapet wall would likely need to be too tall. Consider again the example with an upper level roof width of 150 ft and a balanced snow depth of 1 ft. If the parapet wall height were determined so that it served as an obstruction for the whole upper level (new) roof (that is, h_o tall for something 10 h_o downwind, as discussed in Chapter 3), a 16-ft parapet wall (150 ft/10 + 1.0 ft = 16) would be required. Using the more conservative "sailors' rule of thumb" of 1 to 6, a 26-ft wall (150 ft/6 ft + 1 = 26 ft) would be required. This is not practical.

Approach 4: Baffles

The final approach, shown in **Figure XII-9,** involves a set of baffles at the roof step that redirect wind such that the leeward drift is minimized. Besides potential aesthetic issues, baffles like the parapet wall are effective for leeward drifts only. In addition, a wind tunnel or water flume study would likely be required to establish the size, elevation, and angular orientation of the baffles. Finally, many locations in the United States require a building permit before construction can proceed. It may be difficult to convince the local building official to accept this unique approach since it is not codified.

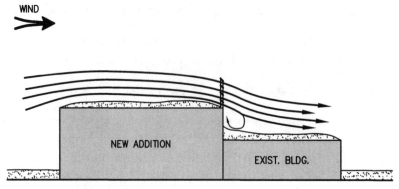

Figure XII-9 Wind Redirection Baffles

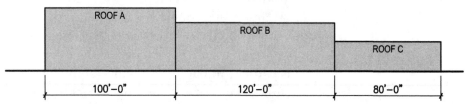

Figure XII-10 Elevation of a Three-Level Roof for Question 7

7. **A building has three nominally flat roof levels, separated by two roof steps as shown in Figure XII-10. Determine the size of the leeward roof step drift on Roof C if the ground snow load is 20 psf.**

The ASCE 7-02 provisions do not address the question of roof steps in series; however, the relationship for drifting at a single roof step can be extended. This extension provides a conservative estimate of the size of the downwind drift on Roof C.

For wind blowing from left to right in **Figure XII-10,** two leeward roof step drifts may form. The drift on Roof B (snow source being Roof A) can be directly calculated from the ASCE 7-02 provisions. For an upwind fetch of 100 ft and a ground load of 20 psf, the maximum surcharge height of the drift on Roof B is

$$(h_d)_B = 0.43 \sqrt[3]{100} \sqrt[4]{20+10} - 1.5$$
$$= 3.17 \text{ ft}$$

Assuming each step is sufficiently large (i.e., none of the drifts is height-constrained), the width of the surcharge is four times its height, and hence the cross-sectional area of the drift is

$$A_B = \frac{1}{2}(3.17 \text{ ft})(4 \times 3.17 \text{ ft}) = 20.1 \text{ ft}^2$$

The leeward drift on Roof C would be caused by snow originally on Roof B and possibly snow originally on Roof A. That is, the upwind fetch for the leeward drift on Roof C is greater than 120 ft but less than 220 ft. It is less than 220 ft because some of the Roof A snow source has been used in forming the drift on Roof B.

If the upwind fetch were 120 ft, the cross-sectional area of the drift on Roof C would be

$$(h_d)_{Cl_u=120\,\text{ft}} = 0.43\sqrt[3]{120}\sqrt[4]{20+10} - 1.5$$
$$= 3.46 \text{ ft}$$

$$(A_C)_{l_u=220\,\text{ft}} = \tfrac{1}{2}(3.46 \text{ ft})(4 \times 3.46 \text{ ft})$$
$$= 23.9 \text{ ft}^2$$

Similarly, if the upwind fetch were 220 ft, the cross-sectional area of the drift on Roof C would be

$$(h_d)_{Cl_u=220\,\text{ft}} = 0.43\sqrt[3]{220}\sqrt[4]{20+10} - 1.5$$
$$= 4.58 \text{ ft}$$

$$(A_C)_{l_u=220\,\text{ft}} = \tfrac{1}{2}(4.58 \text{ ft})(4 \times 4.58 \text{ ft}) = 42.0 \text{ ft}^2$$

It could be argued that the contribution to the drift on Roof C due to the snow on Roof A is simply

$$(A_C)_{\text{roof A}} = 42.0 \text{ ft}^2 - 23.9 \text{ ft}^2 = 18.1 \text{ ft}^2$$

However, not all of the snow that was originally on Roof A can contribute because some was used in forming the drift on Roof B.

The reduced snow source area on Roof A is accounted for by reducing its full contribution of 18.1 ft² by the percentage remaining after the drift on Roof B has been taken into consideration. Recall that in relation to the empirical drift height formula in **Eq. (VII-1),** somewhere between 20% and 50% of snow in the source area typically ended up in the case history drifts. Herein a value of 25% is assumed. Hence, the contribution to the drift on Roof C from the reduced source area of Roof A is

$$(A_C)_{\text{roof A effective}} = 18.1 \text{ ft}^2 (1 - 0.25) = 13.6 \text{ ft}^2$$

Therefore, the total cross-sectional area of the drift on Roof C is the full contribution from Roof B (23.9 ft²) plus the reduced contribution from Roof A (13.6 ft²) or

$$A_C = 23.9 \text{ ft}^2 + 13.6 \text{ ft}^2 = 37.5 \text{ ft}^2$$

Since the roof steps have been assumed sufficiently large (i.e., no height-constrained drifts), the maximum surcharge height for the drift on Roof C is easily calculated from the drift cross-sectional area:

$$\tfrac{1}{2}h_c(4h_c) = 37.5 \ \text{ft}^2$$

or

$$h_c = 4.33 \ \text{ft}$$

Hence, the surcharge height is more than the 3.46-ft height calculated from Roof B alone ($\ell_u = 120$ ft), but less than the 4.58-ft height if both Roofs A and B were fully effective ($\ell_u = 220$ ft). It turns out in this case that the effective fetch is about 195 ft. That is,

$$(h_d)_C = 0.43\sqrt[3]{195}\sqrt[4]{20\ +\ 10} - 1.5$$
$$= 4.33 \ \text{ft}$$

8. A 6-ft-tall snow drift accumulated against a building wall. What lateral pressure would the drift exert against the wall?

As noted in Chapter 9 of this guide, ASCE 7-02 does not address lateral loads exerted by snow. A rough and probably somewhat conservative estimate can be obtained by modeling the snow as a cohesionless soil. According to Rankine's theory for a cohesionless soil and a frictionless wall, the active earth pressure coefficient, K_a (i.e., ratio of lateral to vertical pressure for the soil moving toward the wall) is given by

$$K_a = \frac{1 - \sin\phi}{1 + \sin\phi} \tag{XII-1}$$

where ϕ is the internal friction angle or angle of repose for the soil. This approach for a snowdrift is likely conservative because the lateral pressure caused by a snowdrift against a wall would be less than the pressure caused by a snow mass of infinite horizontal extent being retained by a wall, as assumed in the Rankine theory.

Based on observations, the angle of repose for fresh snow is typically 50 degrees to 60 degrees. After being blown around, however, the snow particles apparently become more rounded (less angular), and the angle of repose is lower. As noted in Chapter 7 (see **Figure VII-4**), for non-full roof step drifts the typical rise to run is 1:4. This suggests an angle of repose of wind-blown snow of about 15 degrees (tan 15° = .268).

A further complication is the fact that the density of snow increases with depth. If the snow density is assumed to increase linearly with depth, which is reasonable in light of **Figure VII-5**, then the lateral pressure would increase as the depth squared. Formally incorporating this variation, however, seems excessive. In the calculation below, the snow density is assumed to be constant and equal to the expected value at the mid-depth of the drift. From **Figure VII-5**, at a depth of 3 ft, the density of the snow is 22 pcf.

From the problem statement, the snow was wind blown. Therefore, the friction angle equals approximately 15 degrees. Thus, the active earth pressure coefficient is

$$K_a = \frac{1 - \sin\phi}{1 + \sin\phi} = \frac{1 - .2588}{1 + .2588} = 0.59$$

The lateral pressure at the bottom of the drift is

$$P_\ell = K_a \, \gamma_{\text{mid-depth}} \, h = 0.59 \ (22 \text{ pcf}) \ (6.0 \text{ ft}) = 78 \text{ psf}$$

while at mid-depth,

$$P_\ell = 0.59 \ (22 \text{ pcf}) \ (3.0 \text{ ft}) = 39 \text{ psf}$$

References

Colbeck, S.C. (1977). "Roof loads resulting from rain-on-snow." *Rep. 77-12,* Cold Regions Research and Engineering Lab., Hanover, N.H.

Davenport, A.G. (1965). "The relationship of wind structure to wind loading." *Proc. symposium on wind effects on buildings and structures, vol. I,* National Physical Lab., Teddington, U.K., pp. 53-102.

National Research Council. (1999). *The impacts of natural disasters: a framework for loss estimation,* National Academy Press, Washington, D.C., p. 68

O'Rourke, M., and Auren, M. (1997). "Snow loads on gable roofs." *J. Struct. Engrg.,* 123(12), pp. 1645-1651.

O'Rourke, M., and DeAngelis, C. (2002). "Snow drift at windward roof steps." *J. Struct. Engrg.,* 128(10), pp. 1330-1336.

O'Rourke, M., and Downey, C. (2001). "Rain-on-snow surcharge for roof design." *J. Struct. Engrg.,* 127(1), pp. 74-79.

O'Rourke, M., Redfield, R., and Von Bradsky, P. (1982). "Uniform snow loads on structures." *J. Struct. Div.,* 108(ST12), pp. 2781-2798.

O'Rourke, M., Koch, P., and Redfield, R. (1983). "Analysis of roof snow load case studies: Uniform loads." *CRREL Rep. 83-1,* U.S. Army Corp. of Engineers, Engineer Research and Development Center, Cold Regions Research and Engineering Lab., Hanover, N.H.

O'Rourke, M., Speck, R., and Stiefel, U. (1985). "Drift snow loads on multi-level roofs." *J. Struct. Engrg.,* 111(2), pp. 290-306.

O'Rourke, M., Tobiasson, W., and Wood, E. (1986). "Proposed code provisions for drifted snow loads." *J. Struct. Engrg.,* 112(9), pp. 2080-2092.

Sachs, P. (1972). *Wind forces in engineering,* Pergamon, Oxford.

Sack, R.L. (1988). "Snow loads on sloped roofs." *J. Struct. Engrg.,* 114(3), pp. 501-517.

Sack, R.L., Burke, G.G., and Penningroth, J. (1984). "Automated data acquisition for structural snow loads." *Proc. 41st Eastern Snow Conf.,* New Carrolton, Md.

SEAW. (1995). *Snow load analysis for Washington,* 2nd ed., Southworth, Wash.

SEAW. (1998). "An analysis of building structural failures, due to the holiday snow storms." Federal Emergency Management Agency, Bothell, Wash.

Speck, R. (1984). "Analysis of snow loads due to drifting on multilevel roofs." MS thesis, Dept. of Civil Engineering, Rensselaer Polytechnic Inst.

Tabler, R. (1994). "Design guidelines for the control of blowing and drifting snow." *Rep. SHRP-H-381,* National Research Council, Strategic Highway Research Program, Washington, D.C.

Tobiasson, W., and Greatorex, A. (1996). "Database and methodology for conducting site specific snow load case studies for the United States." *Proc. 3rd int. conf. on snow engineering,* Sendai, Japan, pp. 249-256.

Tobiasson, W., Buska, J., Greatorex, A., Tirey, J., Fisher, J., and Johnson, S. (2002). "Ground snow loads for New Hampshire." *Tech. Rep. ERDC/ CRREL TR-02-6,* U.S. Army Corp. of Engineers, Engineer Research and Development Center, Cold Regions Research and Engineering Lab., Hanover, N.H.

TopoZone.com

Appendix

This appendix is reproduced from SEI/ASCE Standard 7-02, Second Edition, *Minimum Design Loads for Buildings and Other Structures*, (2003, American Society of Civil Engineers, pp. 77–91).

SECTION 7.0
SNOW LOADS

SECTION 7.1
SYMBOLS AND NOTATIONS

β = gable roof drift parameter as determined from Eq. 7-3

C_e = exposure factor as determined from Table 7-2

C_s = slope factor as determined from Figure 7-2

C_t = thermal factor as determined from Table 7-3

h_b = height of balanced snow load determined by dividing p_f or p_s by γ, in ft (m)

h_c = clear height from top of balanced snow load to (1) closest point on adjacent upper roof, (2) top of parapet, or (3) top of a projection on the roof, in ft (m)

h_d = height of snow drift, in ft (m)

h_e = elevation difference between the ridge line and the eaves

h_o = height of obstruction above the surface of the roof, in ft (m)

I = importance factor as determined from Table 7-4

l_u = length of the roof upwind of the drift, in ft (m)

p_d = maximum intensity of drift surcharge load, in pounds per square ft (kn/m^2)

p_f = snow load on flat roofs ("flat" = roof slope ≤ 5°), in lbs/ft^2 (kn/m^2)

p_g = ground snow load as determined from Figure 7-1 and Table 7-1; or a site-specific analysis, in lbs/ft^2 (kn/m^2)

p_s = sloped roof snow load, in pounds per square ft (kn/m^2)

s = separation distance between buildings, in ft (m)

θ = roof slope on the leeward side, in degrees

w = width of snow drift, in ft (m)

W = horizontal distance from eave to ridge, in ft (m)

γ = snow density, in pounds per cubic ft (kn/m^3) as determined from Eq. 7-4

SECTION 7.2
GROUND SNOW LOADS, p_g

Ground snow loads, p_g, to be used in the determination of design snow loads for roofs shall be as set forth in Figure 7-1 for the contiguous United States and Table 7-1 for Alaska. Site-specific case studies shall be made to determine ground snow loads in areas designated CS in Figure 7-1. Ground snow loads for sites at elevations above the limits indicated in Figure 7-1 and for all sites within the CS areas shall be approved by the authority having

jurisdiction. Ground snow load determination for such sites shall be based on an extreme-value statistical-analysis of data available in the vicinity of the site using a value with a 2% annual probability of being exceeded (50-year mean recurrence interval).

Snow loads are zero for Hawaii, except in mountainous regions as determined by the authority having jurisdiction.

SECTION 7.3
FLAT ROOF SNOW LOADS p_f

The snow load, p_f, on a roof with a slope equal to or less than 5° (1 in./ft = 4.76°) shall be calculated in lbs/ft^2 (kn/m^2) using the following formula:

$$p_f = 0.7 C_e C_t I p_g \qquad \textbf{(Eq. 7-1)}$$

but not less than the following minimum values for low-slope roofs as defined in Section 7.3.4:

where p_g is 20 lb/ft^2 (0.96 kN/m^2) or less,

$\qquad p_f = (I) p_g$ (*Importance factor times p_g*)

where p_g exceeds 20 lb/ft^2 (0.96 kN/m^2),

$\qquad p_f = 20(I)$ (*Importance factor times 20 lb/ft^2*)

7.3.1 Exposure Factor, C_e. The value for C_e shall be determined from Table 7-2.

7.3.2 Thermal Factor, C_t. The value for C_t shall be determined from Table 7-3.

7.3.3 Importance Factor, I. The value for I shall be determined from Table 7-4.

7.3.4 Minimum Values of p_f for Low-Slope Roofs. Minimum values of p_f shall apply to monoslope roofs with slopes less than 15 degrees, hip, and gable roofs with slopes less than or equal to $(70/W) + 0.5$ with W in ft (in SI: $21.3/W + 0.5$, with W in m), and curved roofs where the vertical angle from the eaves to the crown is less than 10 degrees.

SECTION 7.4
SLOPED ROOF SNOW LOADS, p_s

Snow loads acting on a sloping surface shall be assumed to act on the horizontal projection of that surface. The sloped roof snow load, p_s, shall be obtained by multiplying the flat roof snow load, p_f, by the roof slope factor, C_s:

$$p_s = C_s p_f \qquad \text{(Eq. 7-2)}$$

Values of C_s for warm roofs, cold roofs, curved roofs, and multiple roofs are determined from Sections 7.4.1 through 7.4.4. The thermal factor, C_t, from Table 7-3 determines if a roof is "cold" or "warm." "Slippery surface" values shall be used only where the roof's surface is unobstructed and sufficient space is available below the eaves to accept all the sliding snow. A roof shall be considered unobstructed if no objects exist on it that prevent snow on it from sliding. Slippery surfaces shall include metal, slate, glass, and bituminous, rubber, and plastic membranes with a smooth surface. Membranes with an imbedded aggregate or mineral granule surface shall not be considered smooth. Asphalt shingles, wood shingles, and shakes shall not be considered slippery.

7.4.1 Warm Roof Slope Factor, C_s. For warm roofs ($C_t \le 1.0$ as determined from Table 7-3) with an unobstructed slippery surface that will allow snow to slide off the eaves, the roof slope factor C_s shall be determined using the dashed line in Figure 7-2a, provided that for nonventilated warm roofs, their thermal resistance (R-value) equals or exceeds 30 ft²·hr·°F/Btu (5.3 K·m²/W) and for warm ventilated roofs, their R-value equals or exceeds 20 ft²·hr·°F/Btu (3.5 K·m²/W). Exterior air shall be able to circulate freely under a ventilated roof from its eaves to its ridge. For warm roofs that do not meet the aforementioned conditions, the solid line in Figure 7-2a shall be used to determine the roof slope factor C_s.

7.4.2 Cold Roof Slope Factor, C_s. Cold roofs are those with a $C_t > 1.0$ as determined from Table 7-3. For cold roofs with $C_t = 1.1$ and an unobstructed slippery surface that will allow snow to slide off the eaves, the roof slope factor C_s shall be determined using the dashed line in Figure 7-2b. For all other cold roofs with $C_t = 1.1$, the solid line in Figure 7-2b shall be used to determine the roof slope factor C_s. For cold roofs with $C_t = 1.2$ and an unobstructed slippery surface that will allow snow to slide off the eaves, the roof slope factor C_s shall be determined using the dashed line on Figure 7-2c. For all other cold roofs with $C_t = 1.2$, the solid line in Figure 7-2c shall be used to determine the roof slope factor C_s.

7.4.3 Roof Slope Factor for Curved Roofs. Portions of curved roofs having a slope exceeding 70 degrees shall be considered free of snow load, (i.e., $C_s = 0$). Balanced loads shall be determined from the balanced load diagrams in

Figure 7-3 with C_s determined from the appropriate curve in Figure 7-2.

7.4.4 Roof Slope Factor for Multiple Folded Plate, Sawtooth, and Barrel Vault Roofs. Multiple folded plate, sawtooth, or barrel vault roofs shall have a $C_s = 1.0$, with no reduction in snow load because of slope (i.e., $p_s = p_f$).

7.4.5 Ice Dams and Icicles Along Eaves. Two types of warm roofs that drain water over their eaves shall be capable of sustaining a uniformly distributed load of $2p_f$ on all overhanging portions there: those that are unventilated and have an R-value less than 30 ft²·hr·°F/Btu (5.3 k·m²/W) and those that are ventilated and have an R-value less than 20 ft²·hr·°F/Btu (3.5 k·m²/W). No other loads except dead loads shall be present on the roof when this uniformly distributed load is applied.

SECTION 7.5
PARTIAL LOADING

The effect of having selected spans loaded with the balanced snow load and remaining spans loaded with half the balanced snow load shall be investigated as follows:

7.5.1 Continuous Beam Systems. Continuous beam systems shall be investigated for the effects of the three loadings shown in Figure 7-4:

Case 1: Full balanced snow load on either exterior span and half the balanced snow load on all other spans.

Case 2: Half the balanced snow load on either exterior span and full balanced snow load on all other spans.

Case 3: All possible combinations of full balanced snow load on any two adjacent spans and half the balanced snow load on all other spans. For this case there will be $(n - 1)$ possible combinations where n equals the number of spans in the continuous beam system.

If a cantilever is present in any of the above cases, it shall be considered to be a span.

Partial load provisions need not be applied to structural members that span perpendicular to the ridge line in gable roofs with slopes greater than $70/W + 0.5$ with W in ft (in SI: $21.3/W + 0.5$, with W in m).

7.5.2 Other Structural Systems. Areas sustaining only half the balanced snow load shall be chosen so as to produce the greatest effects on members being analyzed.

SECTION 7.6
UNBALANCED ROOF SNOW LOADS

Balanced and unbalanced loads shall be analyzed separately. Winds from all directions shall be accounted for when establishing unbalanced loads.

7.6.1 Unbalanced Snow Loads for Hip and Gable Roofs.

For hip and gable roofs with a slope exceeding $70°$ or with a slope less than $70/W + 0.5$ with W in ft (in SI: $21.3/W + 0.5$, with W in m), unbalanced snow loads are not required to be applied. For roofs with an eave to ridge distance, W, of 20 ft (6.1 m) or less, the structure shall be designed to resist an unbalanced uniform snow load on the leeward side equal to $1.5p_s/C_e$. For roofs with $W > 20$ ft (6.1 m), the structure shall be designed to resist an unbalanced uniform snow load on the leeward side equal to $1.2(1 + \beta/2)p_s/C_e$ with β given by Eq. 7-3.

$$\beta = \begin{cases} 1.0 & p_g \leq 20 \text{ lb/ft}^2 \\ 1.5 - 0.025p_g & 20 < p_g < 40 \text{ lb/ft}^2 \\ 0.5 & p_g \geq 40 \text{ lb/ft}^2 \end{cases}$$
$$\text{(Eq. 7-3)}$$

In SI:

$$\beta = \begin{cases} 1.0 & p_g \leq 0.97 \text{ kN/m}^2 \\ 1.5 - 0.52p_g & 0.97 < p_g < 1.93 \text{ kN/m}^2 \\ 0.5 & p_g \geq 1.93 \text{ kN/m}^2 \end{cases}$$

For the unbalanced situation with $W > 20$ ft (6.1 m), the windward side shall have a uniform load equal to $0.3p_s$. Balanced and unbalanced loading diagrams are presented in Figure 7-5.

7.6.2 Unbalanced Snow Loads for Curved Roofs.

Portions of curved roofs having a slope exceeding 70 degrees shall be considered free of snow load. If the slope of a straight line from the eaves (or the 70-degree point, if present) to the crown is less than 10 degrees or greater than 60 degrees, unbalanced snow loads shall not be taken into account.

Unbalanced loads shall be determined according to the loading diagrams in Figure 7-3. In all cases the windward side shall be considered free of snow. If the ground or another roof abuts a Case II or Case III (see Figure 7-3) curved roof at or within 3 ft (0.91 m) of its eaves, the snow load shall not be decreased between the 30-degree point and the eaves but shall remain constant at the 30-degree point value. This distribution is shown as a dashed line in Figure 7-3.

7.6.3 Unbalanced Snow Loads for Multiple Folded Plate, Sawtooth, and Barrel Vault Roofs.

Unbalanced loads shall be applied to folded plate, sawtooth, and barrel vaulted multiple roofs with a slope exceeding 3/8 in./ft (1.79 degrees). According to 7.4.4, $C_s = 1.0$ for such roofs, and the balanced snow load equals p_f. The unbalanced snow load shall increase from one-half the balanced load at the ridge or crown (i.e., $0.5p_f$) to two times the balanced load given in 7.4.4 divided by C_e at the valley (i.e., $2 p_f/C_e$). Balanced and unbalanced loading diagrams for a sawtooth roof are presented in Figure 7-6. However, the snow surface above the valley shall not be at an elevation higher than the snow above the ridge. Snow depths shall be determined by dividing the snow load by the density of that snow from Eq. 7-4, which is in Section 7.7.1.

7.6.4 Unbalanced Snow Loads for Dome Roofs.

Unbalanced snow loads shall be applied to domes and similar rounded structures. Snow loads, determined in the same manner as for curved roofs in Section 7.6.2, shall be applied to the downwind 90-degree sector in plan view. At both edges of this sector, the load shall decrease linearly to zero over sectors of 22.5 degrees each. There shall be no snow load on the remaining 225-degree upwind sector.

SECTION 7.7
DRIFTS ON LOWER ROOFS
(AERODYNAMIC SHADE)

Roofs shall be designed to sustain localized loads from snow drifts that form in the wind shadow of (1) higher portions of the same structure and, (2) adjacent structures and terrain features.

7.7.1 Lower Roof of a Structure.

Snow that forms drifts comes from a higher roof or, with the wind from the opposite direction, from the roof on which the drift is located. These two kinds of drifts ("leeward" and "windward," respectively) are shown in Figure 7-7. The geometry of the surcharge load due to snow drifting shall be approximated by a triangle as shown in Figure 7-8. Drift loads shall be superimposed on the balanced snow load. If h_c/h_b is less than 0.2, drift loads are not required to be applied.

For leeward drifts, the drift height h_d shall be determined directly from Figure 7-9 using the length of the upper roof. For windward drifts, the drift height shall be determined by substituting the length of the lower roof for l_u in Figure 7-9 and using three-quarters of h_d as determined from Figure 7-9 as the drift height. The larger of these two heights shall be used in design. If this height is equal to or less than h_c, the drift width, w, shall equal $4h_d$ and the drift height shall equal h_d. If this height exceeds h_c, the drift width, w, shall equal $4h_d^2/h_c$ and the drift height shall equal h_c. However, the drift width, w, shall not be greater than $8h_c$. If the drift width, w, exceeds the width of the lower roof, the drift shall be truncated at the far edge of the roof, not reduced to zero there. The maximum intensity of the drift surcharge load, p_d, equals $h_d\gamma$ where snow density, γ, is defined in Eq. 7-4:

$$\gamma = 0.13p_g + 14 \text{ but not more than 30 pcf} \quad \text{(Eq. 7-4)}$$

(in SI: $\gamma = 0.426p_g + 2.2$ but not more than 4.7 kN/m^3)

This density shall also be used to determine h_b by dividing p_f (or p_s) by γ (in SI: also multiply by 102 to get the depth in m).

7.7.2 Adjacent Structures and Terrain Features. The requirements in Section 7.7.1 shall also be used to determine drift loads caused by a higher structure or terrain feature within 20 ft (6.1 m) of a roof. The separation distance, s, between the roof and adjacent structure or terrain feature shall reduce applied drift loads on the lower roof by the factor $(20 - s)/20$ where s is in ft $[(6.1 - s)/6.1$ where s is in m].

SECTION 7.8
ROOF PROJECTIONS

The method in Section 7.7.1 shall be used to calculate drift loads on all sides of roof projections and at parapet walls. The height of such drifts shall be taken as three-quarters the drift height from Figure 7-9 (i.e., $0.75 \, h_d$) with l_u equal to the length of the roof upwind of the projection or parapet wall. If the side of a roof projection is less than 15 ft (4.6 m) long, a drift load is not required to be applied to that side.

SECTION 7.9
SLIDING SNOW

The load caused by snow sliding off a sloped roof onto a lower roof shall be determined for slippery upper roofs with slopes greater than $\frac{1}{4}$ on 12, and for other (i.e., non-slippery) upper roofs with slopes greater than 2 on 12. The total sliding load per unit length of eave shall be $0.4 p_f W$, where W is the horizontal distance from the eave to ridge for the sloped upper roof. The sliding load shall be distributed uniformly on the lower roof over a distance of 15 ft from the upper roof eave. If the width of the lower roof is less than 15 ft, the sliding load shall be reduced proportionally.

The sliding snow load shall not be further reduced unless a portion of the snow on the upper roof is blocked from sliding onto the lower roof by snow already on the lower roof or is expected to slide clear of the lower roof.

Sliding loads shall be superimposed on the balanced snow load.

SECTION 7.10
RAIN-ON-SNOW SURCHARGE LOAD

For locations where p_g is 20 lb/ft^2 (0.96 kN/m^2) or less but not zero, all roofs with a slope less than 1/2 in./ft (2.38°), shall have a 5 lb/ft^2 (0.24 kN/m^2) rain-on-snow surcharge load applied to establish the design snow loads. Where the minimum flat roof design snow load from 7.3.4 exceeds p_f as determined by Eq. 7-1, the rain-on-snow surcharge load shall be reduced by the difference between these two values with a maximum reduction of 5 lb/ft^2 (0.24 kN/m^2).

SECTION 7.11
PONDING INSTABILITY

Roofs shall be designed to preclude ponding instability. For roofs with a slope less than 1/4 in./ft (1.19°), roof deflections caused by full snow loads shall be investigated when determining the likelihood of ponding instability from rain-on-snow or from snow meltwater (see Section 8.4).

SECTION 7.12
EXISTING ROOFS

Existing roofs shall be evaluated for increased snow loads caused by additions or alterations. Owners or agents for owners of an existing lower roof shall be advised of the potential for increased snow loads where a higher roof is constructed within 20 ft (6.1 m). See footnote to Table 7-2 and Section 7.7.2.

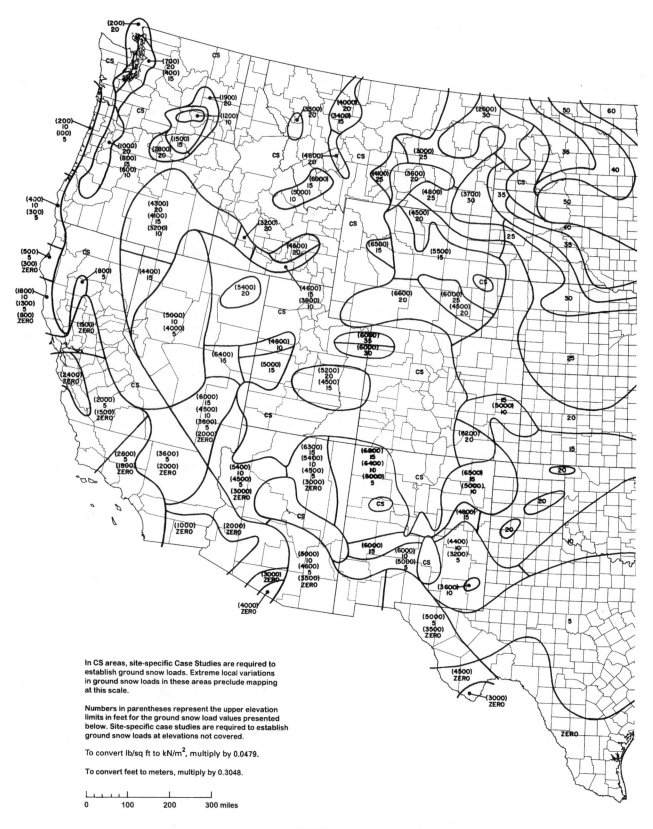

In CS areas, site-specific Case Studies are required to
establish ground snow loads. Extreme local variations
in ground snow loads in these areas preclude mapping
at this scale.

Numbers in parentheses represent the upper elevation
limits in feet for the ground snow load values presented
below. Site-specific case studies are required to establish
ground snow loads at elevations not covered.

To convert lb/sq ft to kN/m^2, multiply by 0.0479.

To convert feet to meters, multiply by 0.3048.

0   100   200   300 miles

FIGURE 7-1
GROUND SNOW LOADS, p_g FOR THE UNITED STATES (IB/SQ FT)

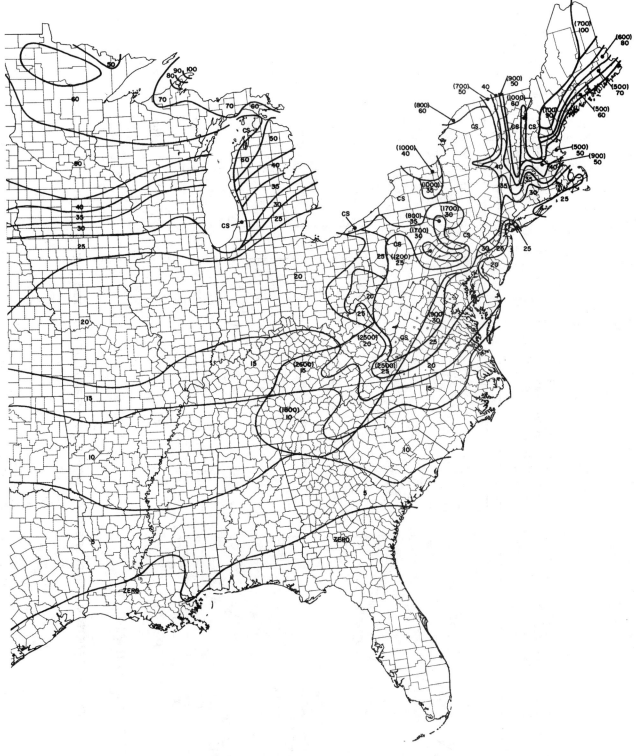

FIGURE 7-1 — continued
GROUND SNOW LOADS, p_g FOR THE UNITED STATES (IB/SQ FT)

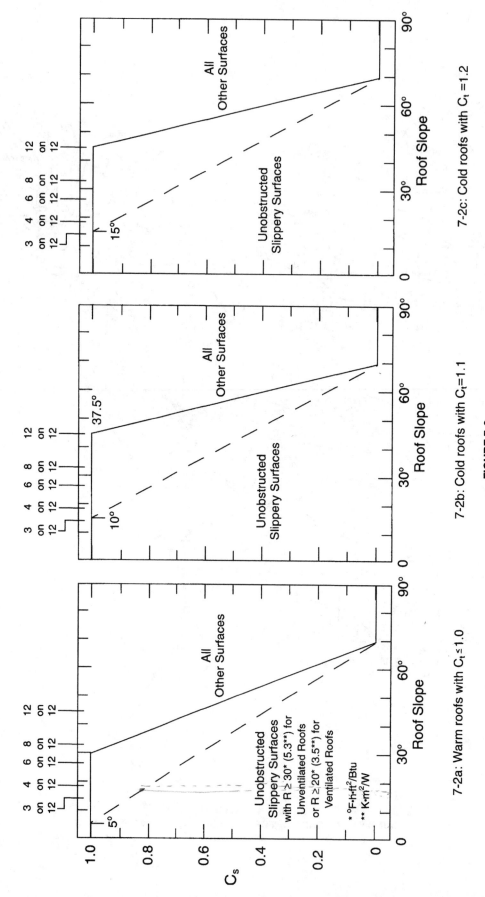

7-2a: Warm roofs with $C_t \leq 1.0$

7-2b: Cold roofs with $C_t = 1.1$

7-2c: Cold roofs with $C_t = 1.2$

FIGURE 7-2

GRAPHS FOR DETERMINING ROOF SLOPE FACTOR C_s FOR WARM AND COLD ROOFS (SEE TABLE 7-3 FOR C_t DEFINITIONS)

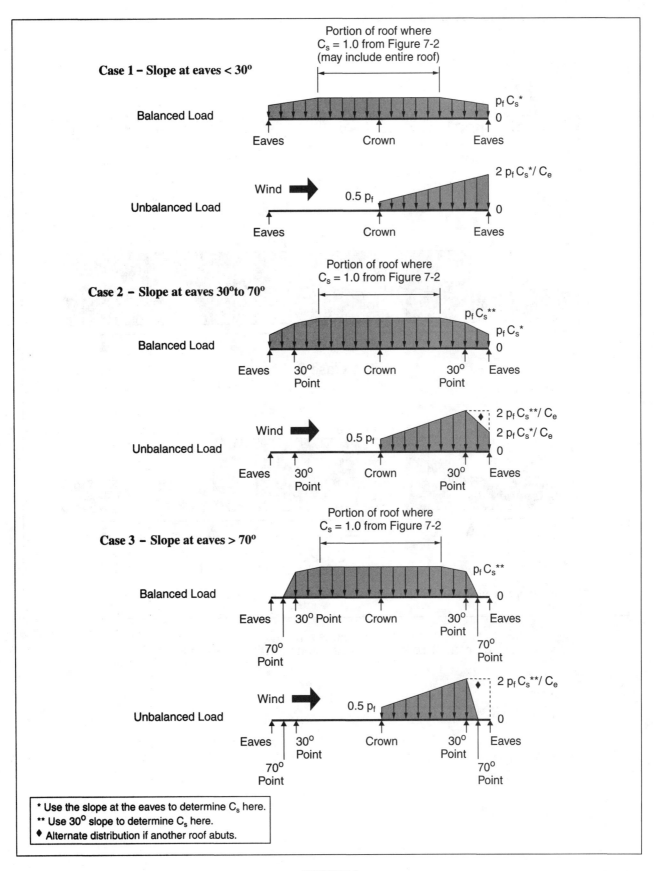

FIGURE 7-3
BALANCED AND UNBALANCED LOADS FOR CURVED ROOFS

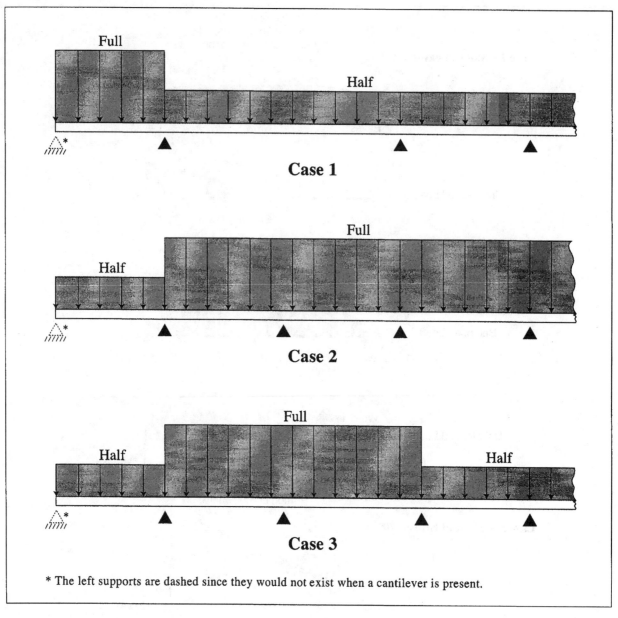

* The left supports are dashed since they would not exist when a cantilever is present.

FIGURE 7-4
PARTIAL LOADING DIAGRAMS FOR CONTINUOUS BEAMS

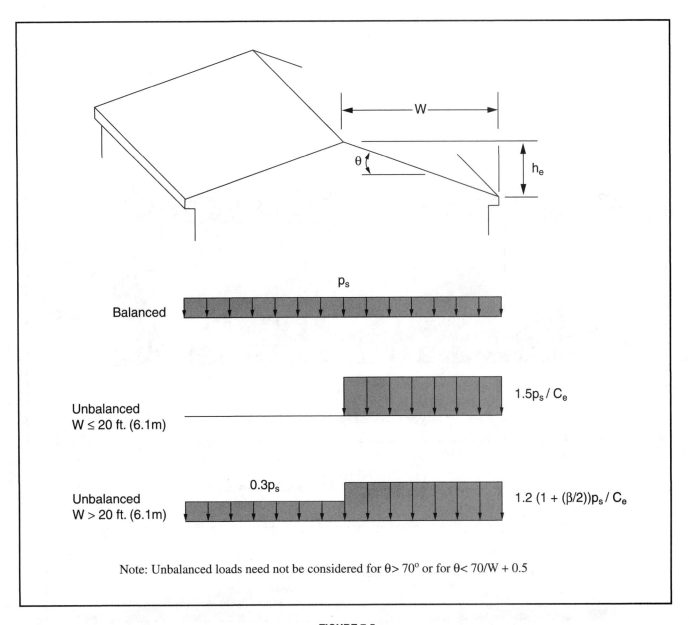

Note: Unbalanced loads need not be considered for $\theta > 70°$ or for $\theta < 70/W + 0.5$

FIGURE 7-5
BALANCED AND UNBALANCED SNOW LOADS FOR HIP AND GABLE ROOFS

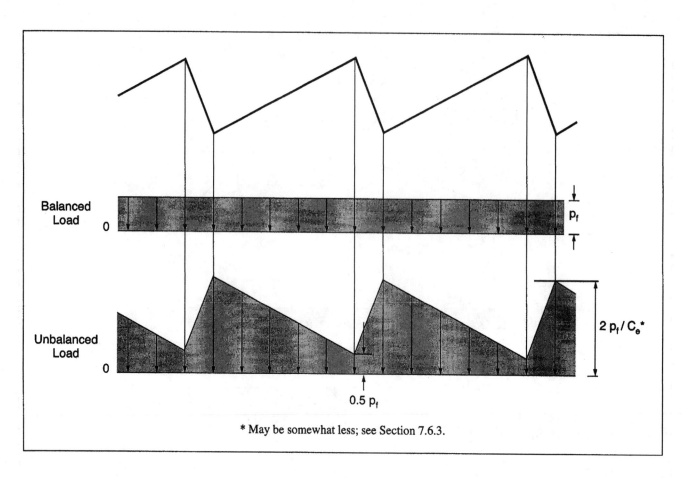

FIGURE 7-6
BALANCED AND UNBALANCED SNOW LOADS FOR A SAWTOOTH ROOF

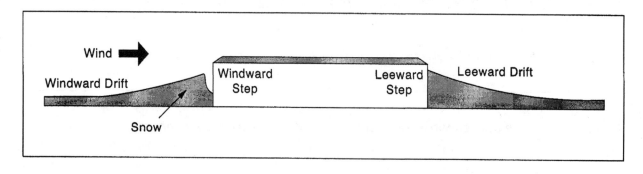

FIGURE 7-7
DRIFTS FORMED AT WINDWARD AND LEEWARD STEPS

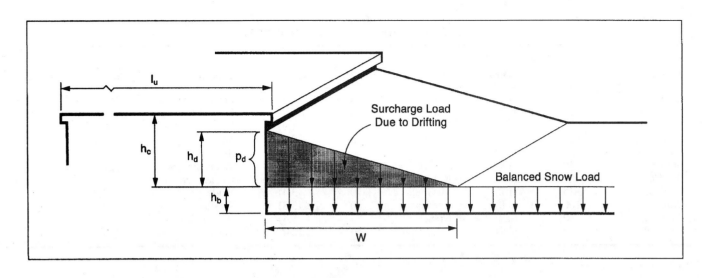

FIGURE 7-8
CONFIGURATION OF SNOW DRIFTS ON LOWER ROOFS

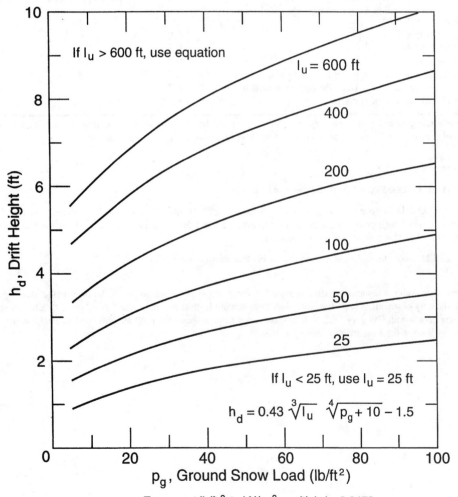

If I_u > 600 ft, use equation

I_u = 600 ft

400

200

100

50

25

If I_u < 25 ft, use I_u = 25 ft

$$h_d = 0.43 \sqrt[3]{I_u} \sqrt[4]{p_g + 10} - 1.5$$

h_d, Drift Height (ft)

p_g, Ground Snow Load (lb/ft^2)

To convert lb/ft^2 to kN/m^2, multiply by 0.0479.
To convert feet to meters, multiply by 0.3048.

FIGURE 7-9
GRAPH AND EQUATION FOR DETERMINING DRIFT HEIGHT, h_d

TABLE 7-1
GROUND SNOW LOADS, p_g, FOR ALASKAN LOCATIONS

Location	p_g lb/ft²	(kN/m²)	Location	p_g lb/ft²	(kN/m²)	Location	p_g lb/ft²	(kN/m²)
Adak	30	(1.4)	Galena	60	(2.9)	Petersburg	150	
Anchorage	50	(2.4)	Gulkana	70	(3.4)	St Paul Islands	40	
Angoon	70	(3.4)	Homer	40	(1.9)	Seward	50	
Barrow	25	(1.2)	Juneau	60	(2.9)	Shemya	25	
Barter Island	35	(1.7)	Kenai	70	(3.4)	Sitka	50	
Bethel	40	(1.9)	Kodiak	30	(1.4)	Talkeetna	120	
Big Delta	50	(2.4)	Kotzebue	60	(2.9)	Unalakleet	50	
Cold Bay	25	(1.2)	McGrath	70	(3.4)	Valdez	160	
Cordova	100	(4.8)	Nenana	80	(3.8)	Whittier	300	
Fairbanks	60	(2.9)	Nome	70	(3.4)	Wrangell	60	
Fort Yukon	60	(2.9)	Palmer	50	(2.4)	Yakutat	150	

TABLE 7-2
EXPOSURE FACTOR, C_e

Terrain Category	Fully Exposed	Exposure of Roof* Partially Exposed	Sheltered
A (see Section 6.5.6)	N/A	1.1	1.3
B (see Section 6.5.6)	0.9	1.0	1.2
C (see Section 6.5.6)	0.9	1.0	1.1
D (see Section 6.5.6)	0.8	0.9	1.0
Above the treeline in windswept mountainous areas.	0.7	0.8	N/A
In Alaska, in areas where trees do not exist within a 2-mile (3 km) radius of the site.	0.7	0.8	N/A

The terrain category and roof exposure condition chosen shall be representative of the anticipated conditions during the life of the structure. An exposure factor shall be determined for each roof of a structure.

*Definitions

PARTIALLY EXPOSED. All roofs except as indicated below.

FULLY EXPOSED. Roofs exposed on all sides with no shelter** afforded by terrain, higher structures, or trees. Roofs that contain several large pieces of mechanical equipment, parapets that extend above the height of the balanced snow load (h_b), or other obstructions are not in this category.

SHELTERED. Roofs located tight in among conifers that qualify as obstructions.

**Obstructions within a distance of $10h_o$ provide "shelter," where h_o is the height of the obstruction above the roof level. If the only obstructions are a few deciduous trees that are leafless in winter, the "fully exposed" category shall be used except for terrain Category "A." Note that these are heights above the roof. Heights used to establish the terrain category in Section 6.5.3 are heights above the ground.

TABLE 7-3
THERMAL FACTOR, C_t

Thermal Condition*	C_t
All structures except as indicated below	1.0
Structures kept just above freezing and others with cold, ventilated roofs in which the thermal resistance (R-value) between the ventilated space and the heated space exceeds 25 F°·hr·sq ft/Btu (4.4 K·m²/W)	1.1
Unheated structures and structures intentionally kept below freezing	1.2
Continuously heated greenhouses** with a roof having a thermal resistance (R-value) less than 2.0 F°·hr·ft²/Btu(0.4 K·m²/W)	0.85

*These conditions shall be representative of the anticipated conditions during winters for the life of the structure.

**Greenhouses with a constantly maintained interior temperature of 50°F (10°C) or more at any point 3 ft above the floor level during winters and having either a maintenance attendant on duty at all times or a temperature alarm system to provide warning in the event of a heating failure.

TABLE 7-4
IMPORTANCE FACTOR,
I, (SNOW LOADS)

Category*	I
I	0.8
II	1.0
III	1.1
IV	1.2

*See Section 1.5 and Table 1-1.

Index